Domestic Central Heating Wiring Systems and Controls

Domestic Central Heating Wiring Systems and Controls

Second edition

Ray Ward

AMSTERDAM • BOSTON • HEIDELBERG • LONDON • NEW YORK • OXFORD
PARIS • SAN DIEGO • SAN FRANCISCO • SINGAPORE • SYDNEY • TOKYO

Newnes is an imprint of Elsevier

Newnes is an imprint of Elsevier
Linacre House, Jordan Hill, Oxford OX2 8DP, UK
30 Corporate Drive, Suite 400, Burlington, MA 01803, USA

First edition 1998
Reprinted 2000, 2001, 2002, 2003 (three times)
Second edition 2005
Reprinted 2006, 2007 (twice)

British Library Cataloguing in Publication Data
A catalogue record for this book is available from the British Library

Library of Congress Cataloging-in-Publication Data
A catalog record for this book is available from the Library of Congress

ISBN: 978-0-7506-6436-3

For information on all Newnes publications
visit our website at www.newnespress.com

Printed and bound in *Great Britain*

07 08 09 10 10 9 8 7 6 5 4

CONTENTS

PREFACE

The purpose of this unique book is to provide a comprehensive reference manual for hundreds of items of heating and control equipment and provide trained engineers with a vitally important resource so that they will be able to take advantage of the huge changes currently taking place in the heating industry. Since this book was first published in 1998 the need to be conversant with energy controls has increased considerably due to the introduction on 1st April 2002, of a revised Part L of the Building Regulations. These regulations provide for a continuing obligation to install efficient heating equipment and controls.

Part L will continue to be revised; new regulations are to come into force on 1st April 2005 and future revisions could stipulate that only high efficiency condensing boilers can be installed from around 2007/2008. Clearly the opportunities for qualified installers will exist as the replacement market expands and this book will prove invaluable in providing the information necessary as systems are changed from 'old' gravity systems to a more efficient fully pumped system with full controls.

These regulations demand that efficiency is taken into account when installing a new heating system or updating existing systems. This will have a major impact on the domestic heating industry and provide untold business opportunities for installers who educate the consumer on energy efficiency in their homes and the benefits to the environment due to the reduction in carbon emissions.

In new and existing properties, all new systems are required to be fully pumped whereas existing systems will require upgrading to meet the new levels of efficiency required when the boiler is replaced. This could be installing room and cylinder thermostats, timers or thermostatic radiator valves.

There will be occasions when a gravity system cannot be upgraded so when installing a new boiler it will be necessary to fit controls that ensure that the boiler does not fire when there is no demand for heat. This is referred to as a 'boiler interlock' in the regulations. This will most likely be a room thermostat, a cylinder thermostat and a motorized valve, e.g. Honeywell 'C' Plan system.

Gradually, gravity hot water systems will diminish whilst high efficiency condensing boilers will become the norm. The range of condensing boilers available to installers is increasing month by month and this is reflected in this revised edition of this book where the number included has greatly increased. How many manufacturers take on the challenge of producing a back-boiler to meet efficiency limits remains to be seen but the replacement market will be immense.

This second edition includes a huge number of additional combination boilers due to the fact that over two-thirds of boilers sold are combi's. Also included are controls that have been developed using the latest technology. These include room thermostats that use radio frequency and so are 'wire-free' and also programmable room thermostats, an option when installing a combination boiler.

Also added to this edition is the SEDBUK rating where this is known. SEDBUK stands for Seasonal Efficiency of Domestic Boilers in the UK and is the standard for measuring the efficiency of boilers. For many of the older boilers no data is available. Some models vary on the SEDBUK rating depending on the output. Some efficiency values are minimal but will cause a boiler to be in a different band. The efficiency ratings are:

- 90% and above — Band A
- 86–90% — Band B
- 82–86% — Band C
- 78–82% — Band D

- 74–78% Band E
- 70–74% Band F
- Below 70% Band G

More information can be obtained from www.boilers.org.uk or www.heatingcontrols.org.uk

It is intended that those persons with the appropriate skills and knowledge to work safely on electrical systems use this book. It is not intended for the do it yourself enthusiast or unskilled homeowner. A reminder is given that only persons registered with CORGI may carry out work on certain aspects of gas appliances and equipment.

ACKNOWLEDGEMENTS

The author wishes to thank Meryl Brooks for her continued support and assistance in compiling this revised edition. Also, the manufacturers, who have continued to provide data and allow their diagrams and illustrations to be reproduced. Thanks also to Richard Hawkes for his diligence in drawing the system wiring diagrams. Finally, the editorial staff at Elsevier, in particular, Rebecca, Hayley and Matthew, for their continual guidance and expert advice, which was greatly appreciated.

1

Guide to use

It is essential that this section is read and understood thoroughly prior to use of the book. All information supplied is believed to be correct, and as such no responsibility can be taken for errors or misuse of information.

General

All equipment is listed in numerical and alphabetical order within its own section, and where items are included elsewhere this is mentioned. An index at the back of the book gives additional information.

Manufacturers and trade names

Over the years manufacturers have merged or been taken over by other companies and every effort has been made to guide the reader to the correct location for information. However some items were manufactured under two names, e.g. Apollo boiler was produced by Myson and Thorn but Myson have since merged with Potterton, therefore a list is given below of where some difficulties may arise:

(a) ACL, Drayton, Grasslin, Invensys, Motortrol, Switchmaster, Tower
(b) Danfoss, Randall
(c) Baxi, Myson, Potterton, Thorn
(d) Satchwell, Sunvic, Terrier
(e) Landis & Gyr, Landis & Staefa, Siemens

Reference to the manufacturers' and trade names directory in Chapter 14 will also help.

Programmers and time switches

These are listed in Chapter 2 in manufacturer order. The first detail is whether the item is electromechanical (driven by a motor) or electronic. Then the setting programme is indicated, i.e. 24 hour, 5/2 day (weekday/weekend), or 7 day. The terms 'Basic' and/or 'Full' are used if the item is a programmer. The term 'Basic' means that the programmer does not have the facility for programming 'central heating only' and would be used, e.g. in a gravity hot water, pumped central heating system. The term 'Full' means that central heating can be selected without hot water such as is required for a normal fully pumped system. Programmers described as 'Basic/Full' have the facility for either option and details are given on how to alter the programmer as required. The maximum number of switching options (on/off), usually per day, are given as well as the current rating of the programmer switch in amps. The rating given will be for a resistive load and a rating for inductive loads may be given in brackets. Dimensions are also given in millimetres and this information can be extremely useful when having to replace an obsolete or unavailable model.

Programmers and time switches with inbuilt or external sensors or thermostats

Wiring and specification details are broadly similar to that given for the room thermostats and programmers.

Wiring system diagrams

Diagrams of all usual systems are included plus those of systems where special requirements may need to be met. All of the full system diagrams are based on the use of a junction box or wiring centre, although it is of course possible to connect wiring into a suitable programmer by following the wiring through. A full list of wiring diagrams included is given at the beginning of Chapter 12. **Important note:** for clarity all earth connections have been omitted but must be made where required.

Cylinder and pipe, room and frost thermostats

The room and frost thermostat details are listed separately to those of the cylinder and pipe thermostats, although the information given is similar. It can be assumed that all thermostats are suitable for 240V unless stated otherwise. The terminal identification is given as follows:

Common	The 'live in' terminal. In the case of a room thermostat, e.g. this would be from the 'heating on' terminal of the programmer in most cases.
Demand	This contact will be 'made' to the common when the thermostat is calling or demanding heat.
Satisfied	This terminal will be 'made' to the common when the thermostat has reached the required temperature or is 'satisfied'.
Neutral	*Room thermostats* – should be wired where shown as this enables the heat anticipator to function and therefore make the thermostat more accurate and sensitive to alteration in temperature fluctuation. *Cylinder thermostats* – required for Potterton PTT1 and PTT2.

Also included are the available scale settings, the dimensions in millimetres, and the current rating of the thermostat contacts.

Motorized valves

Besides providing for wiring details of motorized valves and actuators as below, information is also included regarding port layout of 3-way valves, current rating of auxiliary switch, if fitted, and pipe sizes available.

Where the motorized valve or actuator is of a common type, e.g. 2-port spring return, diverter (3-port priority) or mid-position 3-port, then wiring may be given as *standard colour flex conductors* and these are as follows:

2-port 5-wire (or 4-wire without earth), spring return

Brown	Energize motor, usually to open valve
Blue	Neutral
Green/yellow	Earth (if fitted)
Orange	Live-in for auxiliary switch
Grey	Live-out from auxiliary switch when valve energized

Note: In the 2-wire auxiliary switch the orange and grey leads can be reversed. They may also have the colour coding black-black, white-white, or black-white, depending upon age and manufacturer.

2-port 6-wire (or 5-wire without earth), spring return, (excluding Sunvic SZ1302/2302)

As above with extra:

White	Live-out from auxiliary switch when valve de-energized. Orange and grey must be wired correctly as above. If this wire is spare then it must be made electrically safe.

3-port diverter – priority, spring return

Brown	Energize motor to open closed port (usually energized to open port to central heating)
Blue	Neutral
Green/yellow	Earth (if fitted)

3-port mid-position

Orange	From cylinder thermostat demand and to boiler and pump. Note that pump may need to be wired into boiler if boiler has pump over-run requirement.
White/brown	From programmer 'central heating on' via room thermostat if fitted.

Grey	From cylinder thermostat satisfied and also from 'hot water off' of programmer if possible. Without this second connection then 'central heating only' could not be selected if programmer is of the Full control type.
Blue	Neutral
Green/yellow	Earth (if fitted)

Boilers

The problem with boilers and associated information is deciding which ones should be included and how old. We have attempted to include all boilers that were still in production in 1989 and to date, therefore some boilers over 15 years old may be included, although it is felt that boilers beyond this time are unlikely to be incorporated into an updated system.

Besides wiring, the following information on boilers is included:

(a) Heat exchanger material
(b) Suitability for sealed systems
(c) Whether for fully pumped systems only
(d) Wall or floor mounted, or back boiler unit
(e) SEDBUK rating.

The wiring of standard boilers is usually of two methods. Either a simple switched live, or, in the case of a boiler with pump overrun, a permanent live, switched live and pump live. Some back boiler units may require a permanent live to enable the bulbs on the fire front to work when the boiler is off. The wiring of combination boilers is usually via a voltage-free switch of a time clock.

Ancillary equipment

Brief details of domestic compensator systems, boiler energy controls and similar are given and these are listed at the beginning of the section.

2

Programmers and time switches

ACL FP

Electromechanical 24 hour Full programmer

1	2	3	4	5	6	7	8	9	10	11
○	○	○	○	○	○	○	○	○	○	○
L MAINS	N				HW ON		HW OFF		CH ON	CH OFF

(a) Fully pumped 2 × 2 port motor open/close valves links 4–9, 5–7
(b) Fully pumped 2 × 2 port spring return valves, 1 × 3-way mid-position valve, Satchell Duoflow Switchmaster Midi and Drayton Flowshare link 1–4–9 and 5–7
(c) Tower or ACL Biflo mid-position valve link 1–4–9
(d) Terminal 3 is a spare terminal

Clock module available as a spare.

On/off × 4
H106 × W113 × D65
Switch rating 6A

ACL MP

Electromechanical 24 hour Basic programmer

1	2	3	4	5	6	7	8	9	10	11
○	○	○	○	○	○	○	○	○	○	○
L MAINS	N				HW ON				CH ON	

(a) Link 1–4 and 6–11 for all systems except (b)
(b) Tower or ACL Biflo mid-position valve link 1–4
(c) Terminals 3, 7 and 8 are spare terminals

Clock module available as a spare.

On/off × 4
H106 × W113 × D65
Switch rating 6A

ACL TC

Electromechanical 24 hour time switch

1	2	3	4	5	6	7	8	9	10	11
○	○	○	○	○	○	○	○	○	○	○
L MAINS	N		see note			ON				

(a) Link L–4–6 for 240V control
(b) Link 4–6 for voltage-free switching – input to terminal 4
(c) Terminals, 2, 8, 9, 10 and 11 are spare terminals

Clock module available as a spare.

On/off × 4
H160 × W113 × D65
Switch rating 10A

ACL TC/7

As TC with 7-day clock fitted

ACL LP 111

N	L	1	2	3	4
MAINS		COM	OFF	ON	SPARE

Voltage-free switching unless L–1 linked

Electronic 24 hour time switch

On/off × 2
H93 × W148 × D31
Switch rating 2A (1A)

ACL LP 112

N	L	1	2	3	4
MAINS		HW OFF	CH OFF	HW ON	CH ON

Electronic 24 hour Basic/Full programmer

On/off × 2
H93 × W148 × D31
Switch rating 2A (1A)
Move slider at rear of programmer to G for Basic control or P for Full control

ACL LP 241

N	L	1	2	3	4
MAINS		HW OFF	CH OFF	HW ON	CH ON

Facility for setting hot water and heating at different times in Full mode

Electronic 24 hour Basic/Full programmer

On/off × 2
H93 × W148 × D31
Switch rating 2A (1A)
Move slider at rear of programmer to G for Basic control or P for Full control

ACL LP 522

N	L	1	2	3	4
MAINS		HW OFF	HW OFF	CH ON	CH ON

Facility for setting hot water and heating at different times in Full mode

Electronic 5/2 day Basic/Full programmer

On/off × 2
H93 × W148 × D31
Switch rating 2A (1A)
Move slider at rear of programmer to G for Basic control or P for Full control

ACL LP 711

N	L	1	2	3	4
MAINS		COM	OFF	ON	SPARE

Voltage-free switching unless L–1 linked

Electronic 7 day time switch

On/off × 2
H93 × W148 × D31
Switch rating 2A (1A)

ACL LP 722

N	L	1	2	3	4
MAINS		HW OFF	CH OFF	HW ON	CH ON

Facility for setting hot water and heating at different time from each other every day in Full mode

Electronic 7 day Basic/Full programmer

On/off × 2
H93 × W148 × D31
Switch rating 2A (1A)
Move slider at rear of programmer to G for Basic control or P for Full control

ACL LS 111

N	L	1	2	3	4
MAINS		COM	OFF	ON	SPARE

Voltage-free switching unless L–1 linked

Electronic 24 hour time switch

On/off × 2
H81 × W165 × D44
Switch rating 2A (1A)

ACL LS 112

N	L	1	2	3	4
MAINS		COM	OFF	HW ON	CH ON

Voltage-free switching unless L–1 linked

Electronic 24 hour Basic programmer

On/off × 2
H81 × W165 × D46
Switch rating 2A (1A)

ACL LS 241

N	L	1	2	3	4
MAINS		HW OFF	CH OFF	HW ON	CH ON

Electronic 24 hour Basic/Full programmer

On/off × 2
H87 × W170 × D47
Switch rating 2A (1A)
Turn screw at rear of programmer to G for Basic control or P for Full control

ACL LS 522

N	L	1	2	3	4
MAINS		HW OFF	CH OFF	HW ON	CH ON

Facility for 5/2 day setting

Electronic 24 hour Basic/Full programmer

On/off × 2
H87 × W170 × D47
Switch rating 2A (1A)
Turn screw at rear of programmer to G for Basic control or P for Full control

ACL LS 711

N	L	1	2	3	4
MAINS		COM	OFF	ON	SPARE

Voltage-free switching unless L–1 linked

Electronic 7 day time switch

On/off × 2
H87 × W170 × D47
Switch rating 2A (1A)

ACL LS 722

N	L	1	2	3	4
MAINS		HW OFF	CH OFF	HW ON	CH ON

Facility for setting hot water and heating at different time from each other every day in Full mode

Electronic 7 day Basic/Full programmer

On/off × 2
H87 × W170 × D47
Switch rating 2A (1A)
Turn screw at rear of programmer to G for Basic control or P for Full control

ACL 2000

As Tower T 2000

Barlo EPR1

As ACL LS 522

Crossling Controller

As Landis & Gyr RWB 2

Danfoss 3001

As Horstmann 425 Coronet

Danfoss 3002

As Horstmann 425 Diadem

Danfoss CP 15

E	N	L	1	2	3	4	5	6
○	○	○	○	○	○	○	○	○
	MAINS		HW OFF	CH OFF	HW ON	CH ON	SPARE	

Electronic 24 hour or 5/2 day Basic/Full programmer

On/off × 3
H88 × W135 × D38
Switch rating 3A (1A)

Danfoss CP 75

E	N	L	1	2	3	4	5	6
○	○	○	○	○	○	○	○	○
	MAINS		HW OFF	CH OFF	HW ON	CH ON	SPARE	

Electronic 7 day or 5/2 day Basic/Full programmer

On/off × 3
H88 × W135 × D38
Switch rating 3A (1A)

Danfoss FP 15

E	N	L	1	2	3	4	5	6
○	○	○	○	○	○	○	○	○
	MAINS		HW OFF	CH OFF	HW ON	CH ON	SPARE	

Facility for setting hot water and central heating at different times to each other

Electronic 24 hour or 5/2 day Basic/Full programmer

On/off × 3
H88 × W135 × D38
Switch rating 3A (1A)

Danfoss FP 75

E	N	L	1	2	3	4	5	6
○	○	○	○	○	○	○	○	○
	MAINS		HW OFF	CH OFF	HW ON	CH ON	SPARE	

Facility for setting hot water and central heating at different times to each other

Electronic 24 hour or 5/2 day Basic/Full programmer

On/off × 3
H88 × W135 × D38
Switch rating 3A (1A)

Danfoss MP 15

Electronic 24 hour or 5/2 day Basic programmer

E	N	L	1	2	3	4	5	6
○	○	○	○	○	○	○	○	○
	MAINS		HW OFF	CH OFF	HW ON	CH ON	SPARE	

On/off × 3
H88 × W135 × D38
Switch rating 3A (1A)

Danfoss MP 75

Electronic 7 day or 5/2 day Basic programmer

E	N	L	1	2	3	4	5	6
○	○	○	○	○	○	○	○	○
	MAINS		HW OFF	CH OFF	HW ON	CH ON	SPARE	

On/off × 3
H88 × W135 × D38
Switch rating 3A (1A)

Danfoss TS 15

Electronic 24 hour or 5/2 day time switch

N	L	1	2	3	4
○	○	○	○	○	○
N	L	COM	OFF	SPARE	ON
MAINS					

Voltage-free switching unless L–1 linked

On/off × 3
H88 × W135 × D38
Switch rating 3A (1A)

Danfoss TS 75

Electronic 7 day or 5/2 day time switch

N	L	1	2	3	4
○	○	○	○	○	○
N	L	COM	OFF	SPARE	ON
MAINS					

Voltage-free switching unless L–1 linked

On/off × 3
H88 × W135 × D38
Switch rating 3A (1A)

Danfoss-Randall SET 1E

Electronic 24 hour time switch

E	N	L	1	2	3	4	5	6
○	○	○	○	○	○	○	○	○
	MAINS			SPARE		OFF	COM	ON

Voltage-free switching unless L–5 linked

On/off × 2
H98 × W158 × D36
Switch rating 3A (1A)

Danfoss-Randall SET 2E

Electronic 24 hour Basic programmer

E	N	L	1	2	3	4	5	6
○	○	○	○	○	○	○	○	○
	MAINS		HW ON	COM	HW OFF	CH ON	COM	CH OFF

Voltage-free switching unless L–2–5 linked

On/off × 2
H98 × W158 × D36
Switch rating 3A (1A)

Danfoss-Randall SET 3E

Electronic 24 hour Basic/Full programmer

E	N	L	1	2	3	4	5	6
○	○	○	○	○	○	○	○	○
	MAINS		HW ON	COM	HW OFF	CH ON	COM	CH OFF

Voltage-free switching unless L–2–5 linked

On/off × 2
H98 × W158 × D36
Switch rating 3A (1A)

Danfoss-Randall SET 3M

Electromechanical 24 hour Basic/Full programmer

E	N	L	1	2	3	4	5	6
○	○	○	○	○	○	○	○	○
	MAINS 240 V		HW ON	COM	HW OFF	CH ON	COM	CH OFF

Voltage-free switching unless L–2–5 linked

On/off × 2
H98 × W158 × D63
Switch rating 3A
Fit link supplied for Basic control

Danfoss-Randall FP 975

Electronic 7 day Basic/Full programmer

E	N	L	1	2	3	4	5	6
○	○	○	○	○	○	○	○	○
	MAINS		HW OFF	COM	HW ON	CH OFF	COM	CH ON

Voltage-free switching unless L–2–5 linked
With facility for 5/2 day setting

On/off × 3
H99 × W150 × D42
Switch rating 3A (1A)
Move slider at rear for Basic control

Danfoss Randall TS 975

Electronic 7 day time switch

E	N	L	1	2	3	4	5	6
○	○	○	○	○	○	○	○	○
	MAINS			SPARE		OFF	COM	ON

Voltage-free switching unless L–5 linked
With facility for 5/2 day setting

On/off × 3
H99 × W150 × D42
Switch rating 3A (1A)

Drayton Tempus 1

Electronic 24 hour time switch

E	N	L	1	2	3	4
○	○	○	○	○	○	○
	MAINS		COM	ON	OFF	SPARE

Voltage-free switching unless L–1 linked

On/off × 2
H84 × W140 × D46
Switch rating 3A (1A)

Drayton Tempus 1 MK2

Electronic 24 hour/5/2 day time switch

N	L	1	2	3	4
○	○	○	○	○	○
MAINS		COM	OFF	ON	SPARE

Note wiring is different to original Tempus 1
Voltage-free switching unless L–1 linked

On/off × 3
H99 × W153 × D36
Switch rating 5A (2A)

Drayton Tempus 2

E	N	L	1	2	3	4
○	○	○	○	○	○	○
	MAINS		COM	ON	OFF	SPARE

Voltage-free switching unless L–1 linked
With facility for 5/2 day setting

Electronic 24 hour time switch

On/off × 2
H84 × W140 × D46
Switch rating 3A (1A)

Drayton Tempus 2 MK2

N	L	1	2	3	4
○	○	○	○	○	○
MAINS		COM	OFF	ON	SPARE

Note wiring is different to original Tempus 2
Voltage-free switching unless L–1 linked

Electronic 5/2 day or 7 day time switch

On/off × 3
H99 × W153 × D36
Switch rating 5A (2A)

Drayton Tempus 3

E	N	L	1	2	3	4
○	○	○	○	○	○	○
	MAINS		HW OFF	CH OFF	HW ON	CH ON

Electronic 24 hour Basic/Full programmer

On/off × 2
H84 × W140 × D46
Switch rating 3A (1A)
For Basic control remove plug from rear of programmer

Drayton Tempus 4

E	N	L	1	2	3	4
○	○	○	○	○	○	○
	MAINS		HW OFF	CH OFF	HW ON	CH ON

Facility for 5/2 day setting

Electronic 24 hour Basic/Full programmer

On/off × 2
H84 × W140 × D46
Switch rating 3A (1A)
For Basic control remove plug from rear of programmer

Drayton Tempus 6

N	L	1	2	3	4
○	○	○	○	○	○
MAINS		HW OFF	CH OFF	HW ON	CH ON

Facility for setting hot water and central heating at different times to each other

Electronic 24 hour or 5/2 day Basic/Full programmer

On/Off × 3
H99 × W153 × D36
Switch rating 5A (2A)

Drayton Tempus 7

E	N	L	1	2	3	4
○	○	○	○	○	○	○
	MAINS		HW OFF	CH OFF	HW ON	CH ON

Facility for setting hot water and heating at different times daily in Full mode

Electronic 7 day Basic/Full programmer

On/off × 2
H84 × W140 × D46
Switch rating 3A (1A)
For Basic control remove plug from rear of programmer

Drayton Tempus 7 MK2

Electronic 5/2 day or 7 day Basic/Full Programmer

N	L	1	2	3	4
○	○	○	○	○	○
MAINS		HW OFF	CH OFF	HW ON	CH ON

Facility for setting hot water and central heating at different times to each other

On/off × 3
H99 × W153 × D36
Switch rating 5A (2A)

Eberle 606

Electromechanical 24 hour Basic programmer

E	1	2	3	4	5	6	7
○	○	○	○	○	○	○	○
E	L MAINS	N	HW ON	SPARE	CH ON	SPARE	N

Terminals 2–7 are internally linked

On/off × 2

Eberle 607

Electromechanical 24 hour time switch with pump switch

E	1	2	3	4	5	6	7
○	○	○	○	○	○	○	○
E	L MAINS	N	HW ON	SPARE	CH ON	SPARE	N

Terminals 2–7 are internally linked

On/off × 2

Eberle 608

Electromechanical 24 hour priority programmer

For diagram see Figure 2.13, page 58

On/off × 2

Eberle 609

See Eberle 633

Eberle 610 and 610/15

Electromechanical 24 hour time switch

E	1	2	3	4	5	6	7
○	○	○	○	○	○	○	○
E	L MAINS	N	ON		SPARE		N

Terminals 2–7 are internally linked

On/off × 2

Eberle 633 (supercedes 609)

Electromechanical 24 hour Full programmer

E	1	2	3	4	5	6	7
○	○	○	○	○	○	○	○
E	L MAINS	N	HW ON	HW OFF	CH ON	CH OFF	L

Link L–7 unless used in conjunction with Honeywell V4073 6-wire mid-position valve (with external relay)

On/off × 2
H193 × W105 × D72

Flash 31031 (FP 124)

Electromechanical 24 hour time switch

N	N	L	1
○	○	○	○
N	MAINS		ON

On/off × 36
H84 × W167 × D44
Switch rating 6A

Flash 31032 (FP 224)

Electromechanical 24 hour Basic programmer

N	N	L	1	2
○	○	○	○	○
	MAINS		HW ON	CH ON

On/off × 36
H84 × W167 × D44
Switch rating 6A

Flash 31033 (FP 324)

Electromechanical 24 hour Full programmer

N	N	L	1	2
○	○	○	○	○
N	MAINS		ON	ON

On/off × 36
H84 × W167 × D44
Switch rating 6A

Flash 31731 (FP 17)

7 day version of 31031 with up to 6 on/offs per day

Flash 31731 (FP 27)

7 day version of 31032 with up to 6 on/offs per day

Flash 31733 (FP 37)

7 day version of 31033 with up to 6 on/offs per day

Glow-Worm M2525

Electromechanical 24 hour Basic programmer

1	2	3	4	5
○	○	○	○	○
L	N MAINS	E	CH ON	HW ON

On/off × 2
H118 × W209 × D55

Glow-Worm Mastermind

As Landis & Gyr RWB 2

Grasslin QE 1

See Tower QE 1

Grasslin QE 2

See Tower QE 2

Grasslin Towerchron QM 1

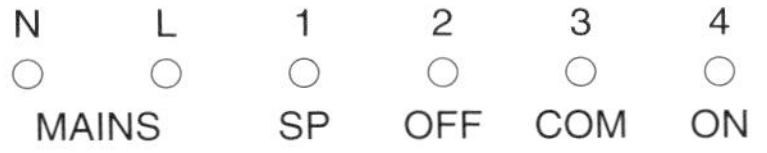

Electromechanical 24 hour time switch

N	L	1	2	3	4
○	○	○	○	○	○
MAINS		SP	OFF	COM	ON

Link L–3 for 240 V switching

On/off × 48
H85 × W156 × D42
Switch rating 5A (2A)

Grasslin Towerchron QM 2

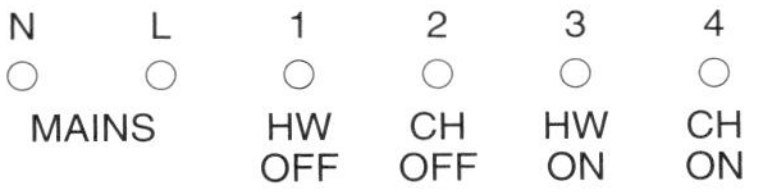

Electromechanical 24 hour Basic/Full Programmer

N	L	1	2	3	4
○	○	○	○	○	○
MAINS		HW OFF	CH OFF	HW ON	CH ON

For Full control remove red pin in back of programmer

On/off × 48
H85 × W156 × D42
Switch rating 5A(2A)

Harp HGC1

Electronic check cost programmer

1	2	3	4	5	6	7	8	E	N	L
○	○	○	○	○	○	○	○	○	○	○
GAS FOR MONITORING	VALVE COST	HW ON	COM	HW OFF	CH ON	COM	CH OFF	E	N MAINS	L

Voltage-free switching unless L–4–7 linked

On/off × 2
H134 × W205 × D48
Switch rating 5A (2A)

Hawk HTC1

See Switchmaster 980

Honeywell ST499A

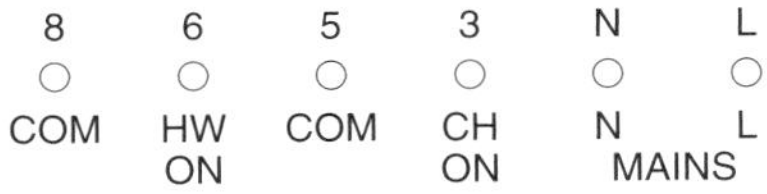

Electronic 24 hour Full programmer

With off/timed/continuous options.

8	6	5	3	N	L
○	○	○	○	○	○
COM	HW ON	COM	CH ON	N MAINS	L

Voltage-free switching unless L–5–8 linked

On/off × 2
H100 × W100 × D38
Switch rating 2A (2A)

Honeywell ST699B

As ST699C, with off/once/twice/continuous options

Honeywell ST699C

Electronic 24 hour Full programmer

8	7	6	5	4	3	N	L
○	○	○	○	○	○	○	○
COM	HW OFF	HW ON	COM	CH OFF	CH ON	N MAINS	L MAINS

Voltage-free switching unless L–5–8 linked

With off/timed/continuous options.

On/off × 2
H100 × W100 × D38
Switch rating 2A (1A)

Honeywell ST799

7 day version of ST699B
If weekday-weekend programming is required, cut and remove link LK1. After replacing programmer press 'Reset' immediately power is turned on.

Honeywell ST6100A

Electronic 24 hour time switch

N	L	1	2	3	4
○	○	○	○	○	○
N MAINS	L MAINS	COM	OFF	SPARE	ON

Voltage-free switching unless L–1 linked

On/off × 3
H95 × W145 × D52
Switch rating 3A (3A)

Honeywell ST6100C

Electronic 7 day time switch

N	L	1	2	3	4
○	○	○	○	○	○
N MAINS	L MAINS	COM	OFF	SPARE	ON

Voltage-free switching unless L–1 linked

On/off × 3
H95 × W145 × D52
Switch rating 3A (3A)

Honeywell ST6200A

Electronic 24 hour Basic programmer

N	L	1	2	3	4
○	○	○	○	○	○
N MAINS	L MAINS	HW OFF	CH OFF	HW ON	CH ON

On/off × 2
H95 × W145 × D52
Switch rating 3A (3A)

Honeywell ST6300A

Electronic 24 hour Full programmer

N	L	1	2	3	4
○	○	○	○	○	○
N MAINS	L MAINS	HW OFF	CH OFF	HW ON	CH ON

On/off × 2
H95 × W145 × D52
Switch rating 3A (3A)

Honeywell ST6400C

Electronic 7 day Full programmer

N	L	1	2	3	4
○	○	○	○	○	○
N	L	HW	CH	HW	CH
MAINS		OFF	OFF	ON	ON

Facility for setting hot water and heating at different times from each other

On/off × 3
H95 × W145 × D52
Switch rating 3A (3A)

Honeywell ST6450

Electronic 5/2 day Full programmer

N	L	1	2	3	4
○	○	○	○	○	○
N	L	HW	CH	HW	CH
MAINS		OFF	OFF	ON	ON

Facility for setting hot water and heating at different times

On/off × 3
H95 × W145 × D52
Switch rating 3A (3A)

Honeywell ST7000A

Electronic 24 hour Basic programmer

CH ON	○	4
HW ON	○	3
HW OFF	○	2
LIVE	○	L

The unit is battery powered and so no neutral is required

On/off × 2
H95 × W122 × D27
Switch rating 2A (2A)

Honeywell ST7000B

Electronic 24 hour time switch

ON	○	3
OFF	○	2
LIVE	○	L

The unit is battery powered and so no neutral is required

On/off × 2
H95 × W122 × D27
Switch rating 2A (2A)

Honeywell ST7100

Electronic 24 hour Full programmer

8	7	6	5	4	3
○	○	○	○	○	○
HW ON	HW OFF	COM	CH ON	CH OFF	COM
				N	L
	○	○	○	○	○
		SPARE		MAINS	

Voltage-free switching unless L–3–6 linked Terminals are provided for earth and neutral connections

Facility for setting hot water and heating at different times to each other during 5/2 day

On/off × 3
H95 × W150 × D49
Switch rating 2A (2A)

Horstmann 423 Amber

Electromechanical 24 hour Full programmer

Designed for use on fully pumped system using change over thermostats and motor open/close motorized valves without end switches.

On/off × 2
H177 × W85 × D57
Switch rating 6A

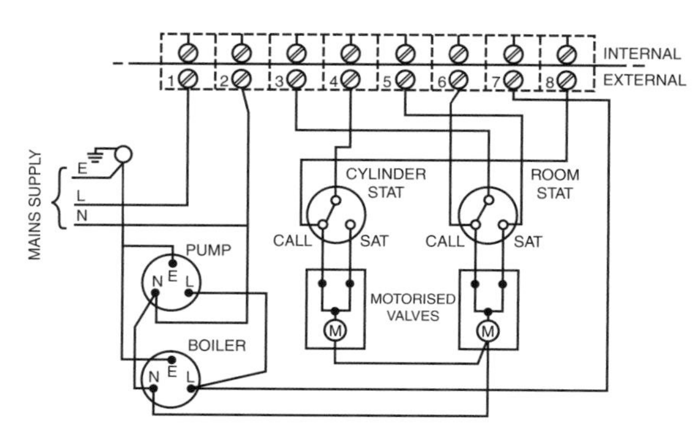

Figure 2.1

Horstmann 423 Amethyst 7+10

Electromechanical 24 hour Full programmer

On/off × 2
H177 × W85 × D57
Switch rating 6A

1	2	3	4	5	6	7	8
○	○	○	○	○	○	○	○
L	N	N	HW	HW	CH	CH	SPARE
MAINS			OFF	ON	OFF	ON	

Amethyst 7 has off/constant/auto control
Amethyst 10 has off/constant/twice/all day control

Horstmann 423 Coral

Electromechanical 24 hour Basic programmer

On/off × 2
H177 × W85 × D57
Switch rating 6A

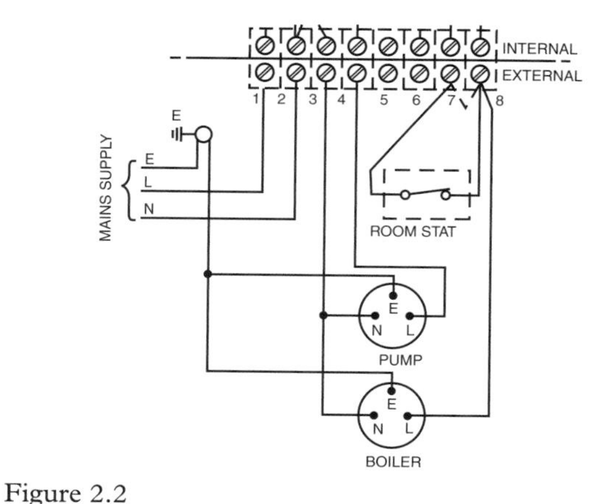

Figure 2.2

If a room thermostat is to be fitted remove link 7–8

Horstmann 423 Diamond

Electromechanical 24 hour Basic programmer

5	N	1	2
○	○	○	○
SPARE	N	L	HW ON
6	L	3	4
○	○	○	○
SPARE	L	L	CH ON

(a) Terminals L–1–3 have a bridging link which can be removed to provide separate switch and motor terminal connections
(b) Terminals 5 and 6 are provided for linking and have no internal connections to the time control

On/off × 2
H105 × W83 × D57
Switch rating 6A (2A)

Horstmann 423 Emerald

Electromechanical 24 hour time switch

N	1	2	5
○	○	○	○
N	SPARE	SPARE	SPARE
L	3	4	6
○	○	○	○
L	COM	ON	SPARE

Terminals L–3 are linked internally but this can be removed for voltage-free switching

On/off × 2
H105 × W83 × D57
Switch rating 6A (2A)

Horstmann 423 Leucite 10

Electromechanical 24 hour Full programmer

1	2	3	4	5	6	7	8
○	○	○	○	○	○	○	○
L	N	HW	HW	COM	CH	COM	CH
MAINS		ON	OFF		ON		OFF

Link 5–7

On/off × 2
H177 × W85 × D57
Switch rating 6A (2A)

Horstmann 423 Pearl 6

Electromechanical 24 hour time switch

N	1	2	5
○	○	○	○
N	SPARE	SPARE	SPARE
L	3	4	6
○	○	○	○
L	COM	ON	SPARE

Terminals L–3 are linked internally but this can be removed for voltage-free switching

On/off × 2
H105 × W88 × D57
Switch rating 6A (2A)

Horstmann 423 Pearl 16

As Pearl 6, but 16A (3A) switch rating

Horstmann 423 Pearl Auto 6 and 16

As Pearl, with off/constant/auto control
See also SMC programmers

Horstmann 423 Ruby

Electromechanical 24 hour time switch

Specifically designed for warm air units.

On/off × 2
H105 × W88 × D57
Switch rating 6A (2A)

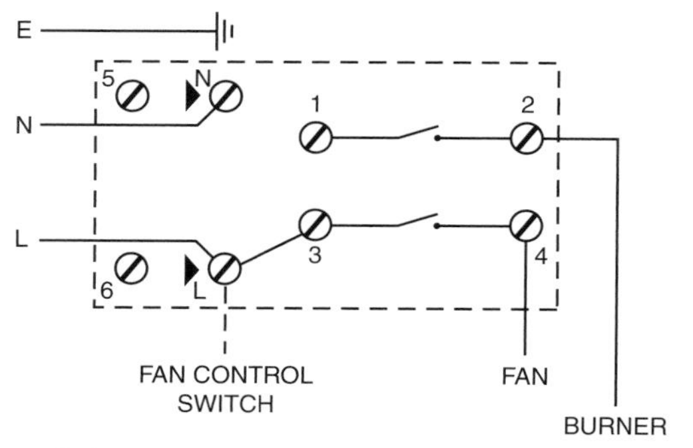

Figure 2.3

Horstmann 423 Sapphire

Electromechanical 24 hour priority programmer

For diagram see Figure 2.14, page 59

On/off × 2
H177 × W85 × D57
Switch rating 6A

Horstmann 423 Topaz

Electromechanical 24 hour time switch

Specifically designed for night set-back thermostat.

On/off × 2
H105 × W88 × D57
Switch rating 6A (2A)

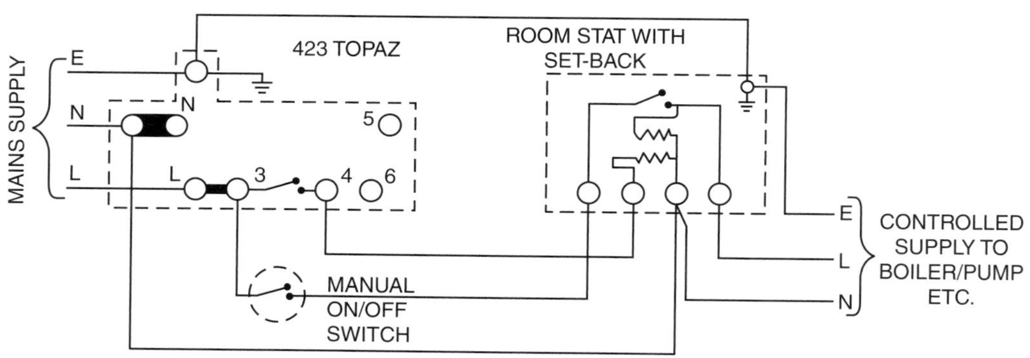

Figure 2.4

Horstmann 424 Amber

Electromechanical 24 hour Full programmer

For diagram see Figure 2.1

Designed for use on fully pumped system, using change over thermostats and motorized valves without end switches.

On/off × 2
H177 × W85 × D57
Switch rating 6A

Horstmann 424 Amethyst 7

Electromechanical 24 hour Full programmer

1	2	3	4	5	6	7	8
○	○	○	○	○	○	○	○
L	N	N	HW OFF	HW ON	CH OFF	CH ON	SPARE
MAINS							

On/off × 2
H177 × W85 × D57
Switch rating 6A

Horstmann 424 Coral

Electromechanical 24 hour Basic programmer

For diagram see Figure 2.2

On/off × 2
H177 × W85 × D57
Switch rating 6A

Horstmann 424 Diamond

Electromechanical 24 hour Basic programmer

		L1		2
		○		○
	MAINS	L		HW ON
	N	3	4	5
	○	○	○	○
MAINS	N	L	CH ON	SPARE

Terminals L–3 are linked internally

On/off × 2
H130 × W87 × D57
Switch rating 6A (2A)

Horstmann 424 Emerald

Electromechanical 24 hour time switch

		L1		2
		○		○
	MAINS	L		SPARE
	N	3	4	5
	○	○	○	○
MAINS	N	COM	ON	SPARE

Terminals L1–3 are linked internally but this can be removed for voltage-free switching

On/off × 2
H130 × W87 × D57
Switch rating 16A (3A)

See also SMC programmers

Horstmann 424 Gem

Electromechanical 24 hour Full programmer

1	2	3	4	5	6	7	8	9	10
○	○	○	○	○	○	○	○	○	○
L	N	N	HW ON	COM	HW OFF	CH ON	COM	CH OFF	L
MAINS									

Link 5–8–10

On/off × 2
H177 × W86 × D57
Switch rating 6A

Horstmann 424 Leucite

Electromechanical 24 hour Full programmer

1	2	3	4	5	6	7	8
○	○	○	○	○	○	○	○
L	N	HW ON	HW OFF	COM	CH ON	COM	CH OFF
MAINS							

Link 5–7

On/off × 2
H177 × W86 × D57
Switch rating 6A

Horstmann 424 Pearl

Electromechanical 24 hour time switch

	L1		2
	○		○
MAINS	L		SPARE

	N	3	4	5
	○	○	○	○
MAINS	N	COM	ON	SPARE

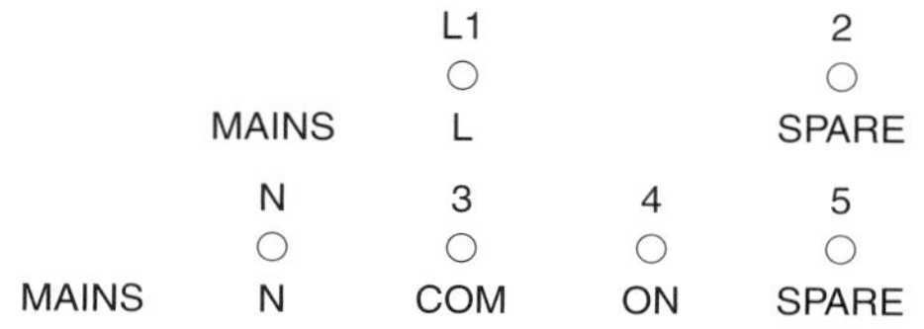

Terminals L1–3 are linked internally but this can be removed for voltage-free switching

On/off × 2
H130 × W87 × D57
Switch rating 16A (3A)

Horstmann 424 Pearl Auto

As Pearl, with off/constant/auto control

Horstmann 424 Sapphire

Electromechanical 24 hour priority programmer

For diagram see Figure 2.14, page 59

On/off × 2
H130 × W87 × D57
Switch rating 6A (3A)

Horstmann 424 Topaz

Electromechanical 24 hour time switch

For diagram see Figure 2.4

Specifically designed for use with night setback thermostat.

On/off × 2
H130 × W87 × D57
Switch rating 6A (3A)

Horstmann 425 Coronet

Electromechanical 24 hour time switch

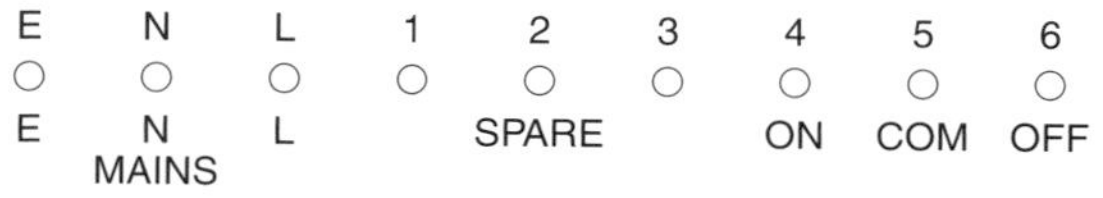

E	N	L	1	2	3	4	5	6
○	○	○	○	○	○	○	○	○
E	N MAINS	L		SPARE		ON	COM	OFF

Voltage-free switching unless L–5 linked

On/off × 2
H107 × W152 × D39
Switch rating 16A (6A)

Horstmann 425 Diadem

Electromechanical 24 hour Basic/Full programmer

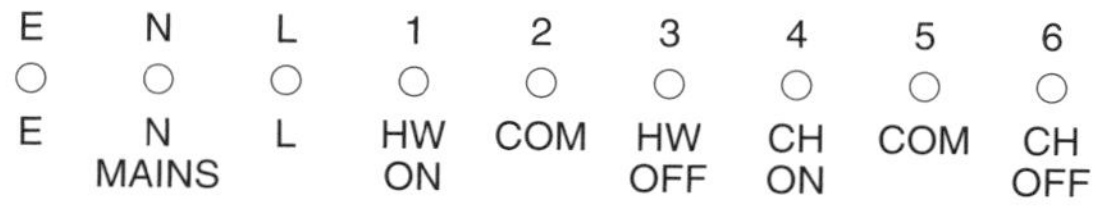

E	N	L	1	2	3	4	5	6
○	○	○	○	○	○	○	○	○
E	N MAINS	L	HW ON	COM	HW OFF	CH ON	COM	CH OFF

Voltage-free switching unless L–2–5 linked

On/off × 2
H107 × W152 × D39
Switch rating 6A (2A)
See 425 TIARA

Horstmann 425 Tiara

As 425 Diadem but without neon indicators

To set programmer for Basic/Full control, turn interlock screws as shown

FULLY PUMPED HOT WATER & CENTRAL HEATING

HW CH

24 hrs
all day
twice
off

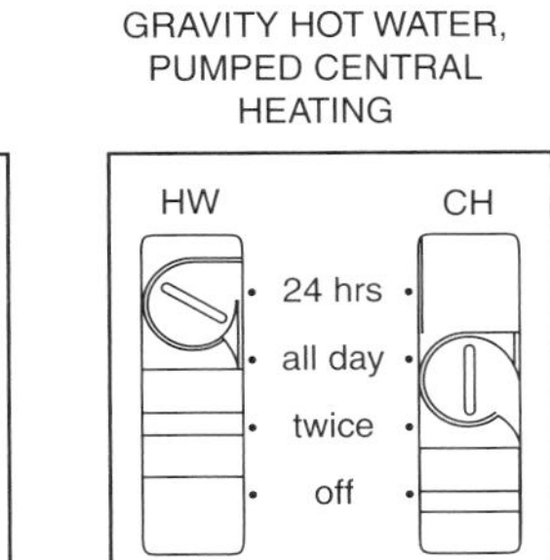

Figure 2.5

Horstmann 517

Electronic 7 day time switch

E	N	L	1	2	3	4	5	6
○	○	○	○	○	○	○	○	○
E	N MAINS	L		SPARE		ON	COM	OFF

Voltage-free switching unless L–5 linked

On/off × 3
H101 × W175 × D45
Switch rating 3A (1A)

Horstmann 525

Electronic 24 hour Basic/Full programmer

E	N	L	1	2	3	4	5	6
○	○	○	○	○	○	○	○	○
E	N MAINS	L	HW ON	COM	HW OFF	CH ON	COM	CH OFF

Voltage-free switching unless L–2–5 linked

Facility for setting hot water and heating at different times to each other daily in Full mode.

On/off × 2
H101 × W175 × D45
Switch rating 3A (1A)

Horstmann 525 7D

For Full mode switch on power to the programmer, remove the switch over plate and move slide switch to extreme left. Move slide switch three positions to the right and re-fit switch cover plate. For Basic mode move the slide switch to the extreme right and fit gravity cover plate and switch on power.

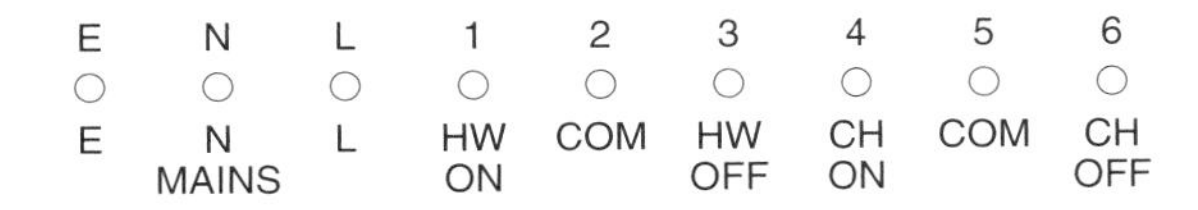

Voltage-free switching unless L–2–5 linked

Electronic 7 day Basic/Full programmer

Facility for setting hot water and heating at different times from each other every day in Full mode.

On/off × 2
H101 × W175 × D45
Switch rating 3A (1A)
Convert to Full/Basic mode as 525

Horstmann 525 Zone

As 527 7D but outputs labelled Zone 1 and Zone 2 instead of HW and CH

Horstmann 581 Senior

Electronic 24 hour time switch

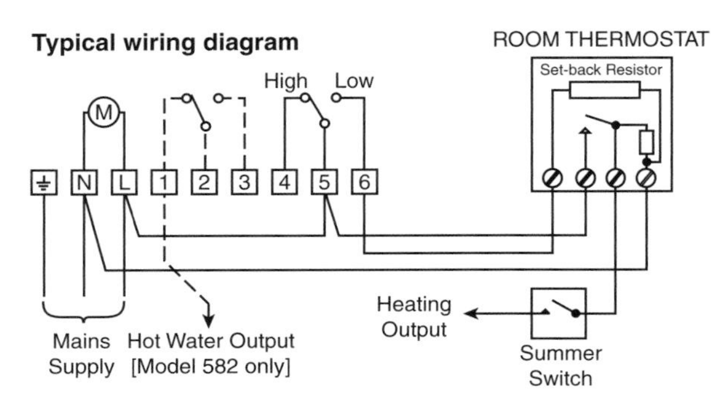

Figure 2.6

Designed for use by the elderly on systems utilizing combination boilers. It has a night-set back facility and is supplied with a suitable room thermostat (Eberle 3545) for which wiring instructions are given. The usual on/off times are featured as high/low.

High/low × 2
H101 × W175 × D45
Switch rating 3A (1A)

Horstmann 582 Senior

Electronic 24 hour Basic/Full programmer

As 581 but with additional hot water switching
See Figure 2.6

On/off × 2
H101 × W175 × D45
Switch rating 3A (1A)
Convert to Full/Basic as 525

Horstmann C 11

Electronic 24 hour time switch

N	L	1	2	3	4
○	○	○	○	○	○
MAINS			COM	OFF	ON

Link L–2 for 240 V control

On/off × 3
H84 × W150 × D29
Switch rating 5A (1A)

Horstmann C 17

Electronic 7 day time switch

7 day version of C11

Horstmann C 21

Electronic 24 hour Basic/Full programmer

N	L	1	2	3	4
○	○	○	○	○	○
MAINS		HW OFF	CH OFF	HW ON	CH ON

Hot water and central heating can be programmed separately

On/off × 3
H84 × W150 × D29
Switch rating 5A (1A)
For basic control remove blue link on back of programmer

Horstmann C 27

Electronic 7 day Basic/Full programmer

Hot water and central heating can be programmed separately for each day of the week.

7 day version of C21

Horstmann C 121

Electronic 24 hour Basic/Full programmer

As C21 except cannot programme hot water and central heating separately

Horstmann C 127

Electronic 7 day Basic/Full programmer

As C27 except cannot programme hot water and central heating separately

Horstmann H 11

Electronic 24 hour time switch

E	N	L	1	2	3	4	5	6
○	○	○	○	○	○	○	○	○
E	N MAINS	L		SPARE		ON	COM	OFF

Voltage-free switching unless L–5 linked

On/off × 3
H101 × W163 × D33
Switch rating 3A (1A)

Horstmann H 17

Electronic 7 day time switch

E	N	L	1	2	3	4	5	6
○	○	○	○	○	○	○	○	○
E	N MAINS	L		SPARE		ON	COM	OFF

Voltage-free switching unless L–5 linked

On/off × 3
H101 × W163 × D33
Switch rating 3A (1A)

Horstmann H 21

Electronic 24 hour Basic/Full programmer

E	N	L	1	2	3	4	5	6
○	○	○	○	○	○	○	○	○
E	N MAINS	L	HW ON	COM	HW OFF	CH ON	COM	CH OFF

Voltage-free switching unless L–2–5 linked

On/off × 3
H101 × W163 × D33
Switch rating 3A (1A)
To change from Basic to Full move slider at rear of programmer

Horstmann H 27

Electronic 7 day Basic/Full programmer

E	N	L	1	2	3	4	5	6
○	○	○	○	○	○	○	○	○
E	N MAINS	L	HW ON	COM	HW OFF	CH ON	COM	CH OFF

Voltage-free switching unless L–2–5 linked

Facility for setting hot water and heating at different times from each other every day in Full mode.

On/off × 3
H101 × W163 × D33
Switch rating 3A (1A)

Horstmann H 27 Z

As H27 but outputs labelled Zone 1 and Zone 2 instead of HW and CH

Horstmann H 37

Electronic 7 day Full programmer

E	N	L						
○	○	○	○	○	○	○	○	○
	MAINS 240V		ON	OFF	ON	OFF	ON	OFF
			ZONE 1		ZONE 2		ZONE 3	

Although the programmer commissioning switch has a gravity position it should not be selected

With one hot water channel and two heating zone channels.

On/off × 3
H101 × W163 × D33
Switch rating 3A (1A)

Horstmann H 121

Electronic 24 hour Basic/Full programmer

E	N	L	1	2	3	4	5	6
○	○	○	○	○	○	○	○	○
E	N MAINS	L	HW ON	COM	HW OFF	CH ON	COM	CH OFF

Voltage-free switching unless L–2–5 linked

On/off × 3
H101 × W163 × D33
Switch rating 3A (1A)

Horstmann SC1 Centaur

Electronic 24 hour time switch

1	○	N
2	○	SPARE
3	○	ON
4	○	LIVE

The unit is battery powered and so no neutral is required

On/off × 3
H71 × W142 × D30
Switch rating 5A (1A)

Horstmann SC 7 Centaur

5/2 day version of SC1 with same wiring and specification

Horstmann TC 1 Centaur

Electronic 24 hour Basic programmer

1	○	N
2	○	CH ON
3	○	HW ON
4	○	LIVE IN

The unit is battery powered and so no neutral is required

On/off × 3
H71 × W142 × D30
Switch rating 5A (1A)

Horstmann TC 7 Centaur

5/2 day version of TC 1 with same wiring and specification

Ideal STD. ISC-1

Electromechanical 24 hour Basic/Full Programmer

N	L	1	2	3	4
○	○	○	○	○	○
N	L	SPARE	CH	SPARE	HW
MAINS			ON		ON

On/off × 2
H105 × W181 × D77
Switch rating 3A (1A)

Invensys SM 1

Electromechanical 24 hour time switch

N	L	1	2	3	4
○	○	○	○	○	○
N	L	COM	OFF	ON	SPARE
MAINS					

Link L–1 for 240 V switching
Internal electronic operation

On/off × 2
H83 × W138 × D55
Switch rating 2A (1A)

Invensys SM 2

Electromechanical 24 hour Basic/Full programmer

N	L	1	2	3	4
○	○	○	○	○	○
N	L	HW	CH	HW	CH
MAINS		OFF	OFF	ON	ON

Internal electronic operation

On/off × 2
H83 × W138 × D55
Switch rating 2A (1A)

Landis & Gyr RWB 1

Electromechanical 24 hour Basic/Full programmer

N	L	3	4
○	○	○	○
N	L	HW	CH
MAINS		ON	ON

On/off × 2
H80 × W135 × D38
Switch rating 10A (2A)
For Full control turn screw at rear of programmer to horizontal

Landis & Gyr RWB 2

Electromechanical 24 hour Basic/Full programmer

N	L	1	2	3	4
○	○	○	○	○	○
N	L	HW	CH	HW	CH
MAINS		OFF	OFF	ON	ON

On/off × 2
H80 × W135 × D38
Switch rating 10A (2A)
For Full control turn screw at rear of programmer to horizontal

Landis & Gyr RWB 2 MK2

As RWB 2 with internal electronic operation.
Switch rating 5A (2A)

Landis & Gyr RWB 2.9

As RWB 2, but without neon indicators

Landis & Staefa RWB 7

Electronic 24 hour, 5/2 day, 7 day time switch

N	L	1	2	3	4
○	○	○	○	○	○
N	L	SPARE	COM	OFF	ON
MAINS					

Voltage-free switching unless L–2 linked

On/off × 2
H85 × W140 × D35
Switch rating 6A (2A)

Landis & Staefa RWB 9

Electronic 24 hour, 5/2 day, 7 day Basic/Full programmer

N	L	1	2	3	4
○	○	○	○	○	○
N	L	HW	CH	HW	CH
MAINS		OFF	OFF	ON	ON

On/off × 2
H85 × W140 × D35
Switch rating 6A (2A)
To change from Full to Basic move dip switch on rear of programmer to '10' position

Landis & Gyr RWB 20

Electronic 7 day Basic/Full programmer

N	L	1	2	3	4
○	○	○	○	○	○
	L	HW	CH	HW	CH
	MAINS	OFF	OFF	ON	ON

The unit is battery powered so no neutral is required. Facility for setting hot water and heating at different times daily in Full mode

On/off × 3
H87 × W135 × D40
Switch rating 6A (2A)
For Full control cut link at rear of programmer

Landis & Gyr RWB 30

Electromechanical 24 hour time switch

N	L	1	2	3	4
○	○	○	○	○	○
N	L	SPARE	COM	OFF	ON
MAINS					

Voltage-free switching unless L–2 linked

On/off × 2
H80 × W135 × D38
Switch rating 6A (2A)

Landis & Gyr RWB 40

Electronic 24 hour Basic/Full programmer

N	L	1	2	3	4
○	○	○	○	○	○
N	L	HW	CH	HW	CH
MAINS		OFF	OFF	ON	ON

On/off × 3
H90 × W115 × D44
Switch rating 6A (2A)
For Full control cut link at rear of programmer

Landis & Gyr RWB 50

Electronic 24 hour time switch

N	L	1	2	3	4
○	○	○	○	○	○
N	L	SPARE	COM	OFF	ON
MAINS					

Voltage-free switching unless L–2 linked

On/off × 3
H90 × W115 × D44
Switch rating 6A (2A)

Landis & Gyr RWB 100

Electronic 24 hour time switch

N	L	1	2	3	4
○	○	○	○	○	○
N	L	SPARE	COM	OFF	ON
MAINS					

Voltage-free switching unless L–2 linked

On/off × 2
H80 × W135 × D31
Switch rating 6A (2A)

Landis & Gyr RWB 102

Electronic 24 hour Basic programmer

N	L	1	2	3	4
○	○	○	○	○	○
N	L	HW OFF	CH OFF	HW ON	CH ON
MAINS					

Note that heating only is not available and no connection need to be made to terminal 1 as, e.g., for a mid-position valve

On/off × 2
H80 × W135 × D31
Switch rating 6A (2A)

Landis & Gyr RWB 152

Electronic 5/2 day time switch

N	L	1	2	3	4
○	○	○	○	○	○
N	L	SPARE	COM	OFF	ON
MAINS					

Voltage-free switching unless L–2 linked

On/off × 2
H80 × W135 × D31
Switch rating 6A (2A)

Landis & Gyr RWB 170

Electronic 7 day time switch

N	L	1	2	3	4
○	○	○	○	○	○
N	L	SPARE	COM	OFF	ON
MAINS					

Voltage-free switching unless L–2 linked

On/off × 2
H80 × W135 × D31
Switch rating 6A (2A)

Landis & Gyr RWB 200

Electronic 24 hour Basic/Full programmer

N	L	1	2	3	4
○	○	○	○	○	○
N	L	HW OFF	CH OFF	HW ON	CH ON
MAINS					

On/off × 2
H80 × W135 × D31
Switch rating 6A (2A)

Landis & Gyr RWB 252

Electronic 5/2 day Basic/Full programmer

N	L	1	2	3	4
○	○	○	○	○	○
N	L	HW OFF	CH OFF	HW ON	CH ON
MAINS					

On/off × 2
H80 × W135 × D31
Switch rating 6A (2A)

Landis & Gyr RWB 270

N	L	1	2	3	4
○	○	○	○	○	○
N	L	HW	CH	HW	CH
MAINS		OFF	OFF	ON	ON

Electronic 7 day programmer

On/off × 2
H80 × W135 × D31
Switch rating 6A (2A)

Landis & Staefa RWB 2E

As RWB 2 with internal electronic operation

Landis & Staefa RWB 30E

As RWB 30 with internal electronic operation

Myson MEP1c

N	L	1	2	3	4
○	○	○	○	○	○
N	L	SPARE	OFF	COM	ON
MAINS					

Link L–3 for 240 V switching

Electronic 24 hour or 5/2 day or 7 day time switch

On/off × 3
H100 × W160
Switch rating 3A (3A)

Myson MEP2c

N	L	1	2	3	4
○	○	○	○	○	○
N	L	HW	CH	HW	CH
MAINS		OFF	OFF	ON	ON

Electronic 24 hour or 5/2 day or 7 day Basic/Full programmer

On/off × 3
H100 × W160
Switch rating 3A (3A)

Myson Microtimer 1

N	○	N MAINS		
L	○	L MAINS		
7	○	HW OFF		
6	○	HW ON	○	9
4	○	CH OFF	○	10
3	○	CH ON	○	11
			SPARE	

A connection block is provided for neutrals

Electronic 24 hour Basic/Full programmer

On/off × 2
H100 × W165 × D50
Switch rating 2A (2A)
To adjust from Full to Basic system, remove programmer from backplate and move system select switch to required position

Myson Microtimer 7

N	○	N MAINS		
L	○	L MAINS		
7	○	HW OFF		
6	○	HW ON	○	9
4	○	CH OFF	○	10
3	○	CH ON	○	11
			SPARE	

A connection block is provided for neutrals

Electronic 7 day Basic/Full programmer

On/off × 2
H100 × W165 × D50
Switch rating 2A (2A)
To adjust from Full to Basic system, remove programmer from backplate and move system select switch to required position

Potterton 423

As Horstmann 423 Diamond

Potterton 424

As Horstmann 424 Diamond

Potterton EP 2000

A	B	C	D	N	L	1	2	3	4	5
○	○	○	○	○	○	○	○	○	○	○
	SPARE			MAINS		HW OFF	CH OFF	HW ON	CH ON	L

Link L–5. A connection block is provided for neutrals and earths

Electronic 24 hour Basic/Full programmer

On/off × 2
H100 × W157 × D46
Switch rating 6A (2A)
To adjust from Basic to Full move slider from 10 to 16 position and turn screw to vertical on rear of programmer

Potterton EP 2000 MK 2

A	B	C	D	N	L	1	2	3	4	5
○	○	○	○	○	○	○	○	○	○	○
	SPARE			MAINS		HW OFF	CH OFF	HW ON	CH ON	L

Link L–5. A connection block is provided for neutrals and earths

Electronic 24 hour Basic/Full programmer

On/off × 2
H104 × W161 × D49
Switch rating 2A (1A)
To adjust from Basic to Full move slider on rear of programmer from 10 to 16 position

Potterton EP 2001

A	B	C	D	N	L	1	2	3	4	5
○	○	○	○	○	○	○	○	○	○	○
	SPARE			MAINS		HW OFF	CH OFF	HW ON	CH ON	L

Link L–5. A connection block is provided for neutrals and earths

Electronic 24 hour Basic/Full programmer

Facility for 5/2 day setting.

On/off × 2
H104 × W160 × D41
Switch rating 6A (2A)
To adjust from Basic to Full move slider from 10 to 16 position on rear of programmer with battery removed

Potterton EP 2002

Electronic 5/2 day Basic/Full programmer

5/2 day version of EP 2000 MK2

Potterton EP 3000

As EP 2000 specification but with 7 day programming facility

Potterton EP 3001

As EP 2001 specification but with 7 day programming facility and allows for hot water and central heating to be set for different times daily

Potterton EP 3002

Electronic 7 day Basic/Full programmer

Facility for setting hot water and heating for different times daily.

7 day version of EP 2000 MK2

Potterton EP 4000

A	B	C	D	N	L	1	2	3	4	5
○	○	○	○	○	○	○	○	○	○	○
	SPARE			MAINS		SP	OFF	SP	ON	COM

Voltage-free switching unless L–5 linked. A connection block is provided for neutrals and earths

Electronic 7 day time switch

On/off × 2
H100 × W157 × D46
Switch rating 6A (2A)

Potterton EP 4000 MK2

A	B	C	D	N	L	1	2	3	4	5
○	○	○	○	○	○	○	○	○	○	○
	SPARE			MAINS		SP	OFF	SP	ON	COM

Voltage-free switching unless L–5 linked. A connection block is provided for neutrals and earths

Electronic 24 hour time switch

On/off × 2
H104 × W161 × D49
Switch rating 2A (1A)

Potterton EP 4001

A	B	C	D	N	L	1	2	3	4	5
○	○	○	○	○	○	○	○	○	○	○
	SPARE			MAINS		SP	OFF	SP	ON	COM

Voltage-free switching unless L–5 linked. A connection block is provided for neutrals and earths

Electronic 5/2 day time switch

On/off × 3
H104 × W160 × D41
Switch rating 6A (2A)

Potterton EP 4002

Electronic 5/2 day time switch

5/2 day version of EP 4000 MK2 with 3 on/offs per day

Potterton EP 5001

A	B	C	D	N	L	1	2	3	4	5
○	○	○	○	○	○	○	○	○	○	○
	SPARE			MAINS		SP	OFF	SP	ON	COM

Voltage-free switching unless L–5 linked. A connection block is provided for neutrals and earths

Electronic 7 day time switch

On/off × 3
H104 × W160 × D41
Switch rating 6A (2A)

Potterton EP 5002

Electronic 7 day time switch

7 day version of EP 4000 MK2 with 3 on/offs per day

Potterton EP 6000

Electronic 7 day Basic/Full programmer

A	B	C	D	N	L	1	2	3	4	5
○	○	○	○	○	○	○	○	○	○	○
	SPARE			MAINS		HW OFF	CH OFF	HW ON	CH ON	L

Link L–5. A connection block is provided for neutrals and earths

Facility for setting either hot water or heating at different times daily and the other channel 5/2 day in Full mode.

On/off × 3
H104 × W164 × D51
Switch rating 6A (2A)
To adjust from Basic to Full move slider from 10 to 16 position on rear of programmer with battery removed

Potterton EP 6002

Electronic 7 day Full programmer

A	B	C	D	N	L	1	2	3	4	5
○	○	○	○	○	○	○	○	○	○	○
	SPARE			MAINS		HW OFF	CH OFF	HW ON	CH ON	L

Link L–5. A connection block is provided for neutrals and earths

Facility for setting hot water and heating for different times to each other daily.

On/off × 3
H104 × W161 × D49
Switch rating 2A (1A)

Potterton Mini-Minder

As Landis & Gyr RWB 2

Potterton Mini-Minder E

Electronic 24 hour Basic/Full programmer

N	L	1	2	3	4
○	○	○	○	○	○
MAINS 240V		HW OFF	CH OFF	HW ON	CH ON

Link L–5. A connection block is provided for neutrals and earths

On/off × 2
H105 × W164 × D51
Switch rating 2A (1A)
To change Basic/Full set both sliders to OFF and turn selector at rear to required position

Potterton Mini-Minder Es

Electronic 24 hour time switch

N	L	1	2	3	4
○	○	○	○	○	○
MAINS 240V		SPARE	COM	OFF	ON

Voltage-free switching unless L–2 linked

On/off × 2
H105 × W164 × D51
Switch rating 2A (1A)

Proheat FP 1

As Flash 31031

Proheat FP 2

As Flash 31032

Proheat FP 3

As Flash 31033

Randall Mk. 1

L	N	E	OUT	IN	L	N	L	N	A	B
○	○	○	○	○	○	○	○	○	○	○
	MAINS		ROOM STAT		PUMP		BOILER		SPARE	

Electromechanical 24 hour Basic programmer

On/off × 2
H100 × W200 × D68

Randall Mk. 2 R6

1	2	3	4	5	6	7
○	○	○	○	○	○	○
L	L	N	HW ON	CH ON	SPARE	DO NOT USE

Link 1–2

Electromechanical 24 hour Basic programmer

On/off × 2
H216 × W102 × D57
Switch rating 5A

Randall 102

1	2	3	E	5	6
○	○	○	○	○	○
HW ON	CH ON	COM	E	N MAINS	L

Voltage-free switching unless 3–6 linked

Electromechanical 24 hour Basic programmer

On/off × 2
H135 × W112 × D69
Switch Rating 6A

Randall 102 E

1	2	3	E	5	6
○	○	○	○	○	○
HW ON	CH ON	COM	E	N MAINS	L

Voltage-free switching unless 3–6 linked

Electronic 24 hour Basic programmer

On/off × 6
H136 × W102 × D47
Switch rating 3A

Randall 102 E5

1	2	3	E	5	6
○	○	○	○	○	○
HW ON	CH ON	COM	E	N MAINS	L

Voltage-free switching unless 3–6 linked

Electronic 5/2 day Basic programmer

On/off × 3
H136 × W102 × D47
Switch rating 3A

Randall 102 E7

Electronic 7 day Basic programmer

1	2	3	E	5	6
○	○	○	○	○	○
HW ON	CH ON	COM	E	N MAINS	L

Voltage-free switching unless 3–6 linked

On/off × 3
H136 × W102 × D47
Switch rating 3A

Randall 103

Electromechanical 24 hour time switch

1	2	3	E	5	6
○	○	○	○	○	○
ON	SPARE	COM	E	N MAINS	L

Voltage-free switching unless 3–6 linked

On/off × 2
H135 × W112 × D69
Switch rating 6A

Randall 103 E

Electronic 24 hour time switch

1	2	3	E	5	6
○	○	○	○	○	○
ON	SPARE	COM	E	N MAINS	L

Voltage-free switching unless 3–6 linked

On/off × 6
H136 × W102 × D47
Switch rating 3A

Randall 103 E5

Electronic 5/2 day time switch

1	2	3	E	5	6
○	○	○	○	○	○
ON	SPARE	COM	E	N MAINS	L

Voltage-free switching unless 3–6 linked

On/off × 3
H136 × W102 × D47
Switch rating 3A

Randall 103 E7

Electronic 7 day time switch

1	2	3	E	5	6
○	○	○	○	○	○
ON	SPARE	COM	E	N MAINS	L

Voltage-free switching unless 3–6 linked

On/off × 3
H136 × W102 × D47
Switch rating 3A

Randall 105

Electromechanical 24 hour Basic programmer

For use with the ACL/Tower Biflo mid-position valve.

For diagram see Figure 2.24, page 61

On/off × 2
H135 × W112 × D69
Switch rating 10A

Randall 106

For diagram see Figure 2.15, page 59

Electromechanical 24 hour priority programmer

On/off × 2
H135 × W112 × D69
Switch rating 10A

Randall 151

1	2	3	E	5	6
○	○	○	○	○	○
ON	OFF	COM	E	N MAINS	L

Voltage-free switching unless 3–6 linked

Electromechanical 24 hour time switch

On/off × 2
H135 × W112 × D69
Switch rating 15A

Randall 152 E

1	2	3	E	5	6
○	○	○	○	○	○
ON	OFF	COM	E	N MAINS	L

Voltage-free switching unless 3–6 linked

Electronic 24 hour time switch

On/off × 6
H136 × W102 × D47
Switch rating 8A

Randall 152 E7

1	2	3	E	5	6
○	○	○	○	○	○
ON	OFF	COM	E	N MAINS	L

Voltage-free switching unless 3–6 linked

Electronic 7 day time switch

On/off × 3
H136 × W102 × D47
Switch rating 8A

Randall 153 E

1	2	3	E	5	6
○	○	○	○	○	○
ON	SPARE	COM	E	N MAINS	L

Voltage-free switching unless 3–6 linked

Electronic 24 hour time switch

On/off × 6
H136 × W102 × D47
Switch rating 15A (4A)

Randall 153 E7

1	2	3	E	5	6
○	○	○	○	○	○
ON	SPARE	COM	E	N MAINS	L

Voltage-free switching unless 3–6 linked

Electronic 7 day time switch

On/off × 3
H136 × W102 × D47
Switch rating 15A (4A)

Randall 701

1	2	3	4	5	6	L	N	E
CH ON	CH OFF	HW ON	HW OFF	L CH	L HW		MAINS	

Voltage-free switching unless 5–6–L linked

Electronic 24 hour Basic programmer

On/off × 3
H108 × W221 × D51
Switch rating 3A

Randall 702

1	2	3	4	5	6	L	N	E
CH ON	CH OFF	HW ON	HW OFF	L CH	L HW		MAINS	

Voltage-free switching unless 5–6–L linked

Electronic 24 hour Full programmer

On/off × 3
H108 × W221 × D51
Switch rating 3A

Randall 811, 841, 842, 851, 852

A range of electronic time switches and programmers intended for commercial use. However, as it is possible to use these in a domestic situation, basic wiring and description are given here.

All models:
On/off × up to 200 on/off operations in any 7 day period
H112 × W226 × D55

Randall 811

	N	L	E	1	2
SPARE		MAINS		COM	ON

Voltage-free switching unless L–1 linked

7 day time switch

With 30A SPST switch

Randall 841

	N	L	E	1	2
SPARE		MAINS		COM	ON

Voltage-free switching unless L–1 linked

Single channel pulsed output time switch

For applications where a time pulsed signal is required to activate equipment. For example, for timed bell ringing to school class changes, or shift changes and breaks in factories.

Randall 842

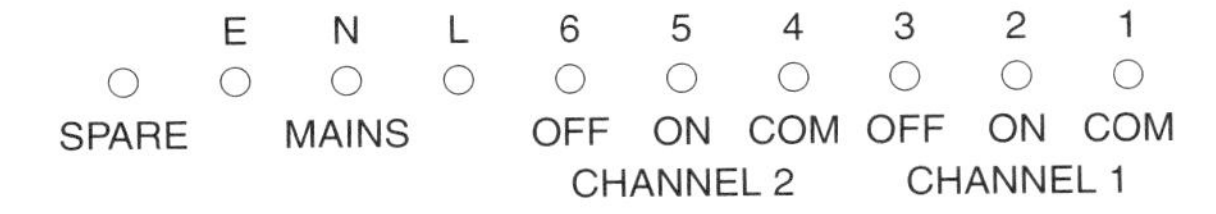

Voltage-free switching unless L–4–1 linked

2-channel pulsed output time switch

Each output channel is totally independant of the other and can be used where timed pulsed signals are required to activate equipment in two separate areas, e.g. in locations as described above.

Randall 851

7 day time switch

With 10A SPDT switch

	E	N	L	6	5	4	3	2	1
○	○	○	○	○	○	○	○	○	○
SPARE		MAINS			SPARE		OFF	ON	COM

Voltage-free switching unless L–1 linked

Randall 852

2 channel 7 day time switch

With independent SPDT switching to each channel

	E	N	L	6	5	4	3	2	1
○	○	○	○	○	○	○	○	○	○
SPARE		MAINS		OFF	ON	COM	OFF	ON	COM
					CHANNEL 2			CHANNEL 1	

Voltage-free switching unless L–4–1 linked

Randall 911

Electronic 24 hour time switch

On/off × 6
H85 × W160 × D38
Switch rating 3A

E	N	L	1	2	3	4	5	6
○	○	○	○	○	○	○	○	○
	MAINS			SPARE		OFF	COM	ON

Voltage-free switching unless L–5 linked

Randall 922

Electronic 24 hour Basic/Full programmer

On/off × 6
H85 × W160 × D38
Switch rating 3A
For basic system the recessed switch on the rear of the programmer should be moved to the **up** position

E	N	L	1	2	3	4	5	6
○	○	○	○	○	○	○	○	○
	MAINS		HW OFF	COM	HW ON	CH OFF	COM	CH ON

Voltage-free switching unless L–2–5 linked

Facility for setting hot water and heating at different times from each other daily in Full mode.

Randall 971

Electronic 7 day time switch

On/off × 3
H85 × W160 × D38
Switch rating 3A

E	N	L	1	2	3	4	5	6
○	○	○	○	○	○	○	○	○
	MAINS			SPARE		OFF	COM	ON

Voltage-free switching unless L–5 linked

Randall 972

Electronic 7 day Basic/Full programmer

E	N	L	1	2	3	4	5	6
○	○	○	○	○	○	○	○	○
	MAINS		HW OFF	COM	HW ON	CH OFF	COM	CH ON

Voltage-free switching unless L–2–5 linked

Facility for setting hot water and heating at different times from each other daily in Full mode.

On/off × 3
H85 × W160 × D38
Switch rating 3A

Randall 3020P

Electromechanical 24 hour Basic programmer

1	2	3	4	5	6	7	E
○	○	○	○	○	○	○	○
N	CH ON	SP	HW ON	SP	L	N MAINS	E

On/off × 2
H216 × W102 × D57
Switch rating 3A

Randall 3022

Electromechanical 24 hour priority programmer

For diagram see Figure 2.17, page 59

On/off × 2
H216 × W102 × D57
Switch rating 3A

Randall 3033

Electromechanical 24 hour Full programmer

1	2	3	4	5	6	7	E
○	○	○	○	○	○	○	○
N	CH ON	CH OFF	HW ON	HW OFF	L	N MAINS	E

On/off × 2
H216 × W102 × D57
Switch rating 3A

Randall 3060

Electromechanical 24 hour Basic programmer

1	2	3	4	5	6	7	E
○	○	○	○	○	○	○	○
N	CH ON	SP	HW ON	SP	L	N MAINS	E

On/off × 2
H216 × W102 × D57
Switch rating 3A

Randall 4033

Electromechanical 24 hour Full programmer

1	2	3	4	5	6	7	E
○	○	○	○	○	○	○	○
L HW	CH ON	CH OFF	HW ON	HW OFF	L	N MAINS	E

Link 1–6

On/off × 2
H216 × W102 × D57
Switch rating 3A

Randall Set 1

E	N	L	1	2	3	4	5	6
	MAINS			SPARE		ON	COM	OFF

Voltage-free switching unless L–5 linked

Electronic 24 hour time switch

On/off × 2
H101 × W148 × D36
Switch rating 5A

Randall Set 2

E	N	L	1	2	3	4	5	6
	MAINS		HW ON	COM	HW OFF	CH ON	COM	CH OFF

Voltage-free switching unless L–2–5 linked

Electronic 24 hour Basic programmer

On/off × 2
H101 × W148 × D36
Switch rating 3A

Randall Set 3

E	N	L	1	2	3	4	5	6
	MAINS		HW ON	COM	HW OFF	CH ON	COM	CH OFF

Voltage-free switching unless L–2–5 linked

Electronic 24 hour Full programmer

On/off × 2
H101 × W148 × D36
Switch rating 3A

Randal Set 1E, 2E, 3E, 3M

See Danfoss Randall 1E, 2E, 3E, 3M

Randall Set 4

E	N	L	1	2	3	4	5	6
	MAINS			SPARE		ON	COM	OFF

Voltage-free switching unless L–5 linked

Electronic 7 day time switch

On/off × 2
H101 × W148 × D36
Switch rating 5A

Randall Set 5

E	N	L	1	2	3	4	5	6
	MAINS		HW ON	COM	HW OFF	CH ON	COM	CH OFF

Voltage-free switching unless L–2–5 linked

Electronic 5/2 day Full programmer

On/off × 2
H101 × W148 × D36
Switch rating 3A

Randall TSR/2

1	2	3	4	5	6	7
L	L	N	SPARE	ON	SPARE	

Link 1–2

Electromechanical 24 hour time switch

On/off × 2
H216 × W102 × D57
Switch rating 3A

Randall TSR 2+2

Electromechanical 24 hour priority programmer

On/off × 2
H216 × W102 × D57
Switch rating 3A

For diagram see Figure 2.16, page 59

Randall TSR 2P

Electromechanical 24 hour Basic programmer

On/off × 2
H216 × W102 × D57
Switch rating 3A

1	2	3	4	5	6	7
○	○	○	○	○	○	○
L	L	N	SPARE	HW ON		CH ON

Link 1–2 and 5–6

Randall TSR 3+3

Electromechanical 24 hour Full programmer

On/off × 2
H216 × W102 × D57
Switch rating 3A

1	2	3	4	5	6	7
○	○	○	○	○	○	○
HW ON	HW OFF	N	CH ON	CH OFF	N MAINS	L MAINS

Ravenheat

As Switchmaster 950

Sangamo M5

Electromechanical 24 hour Basic programmer

On/off × 2
H89 × W141 × D38
Switch rating 10A (2A)

8	○	CH ON
7	○	
6	○	CH L
5	○	N
4	○	N MAINS
3	○	L MAINS
2	○	HW OFF
1	○	HW ON

Link 1–6

Sangamo M6

Electromechanical 24 hour time switch

On/off × 2
H89 × W141 × D38
Switch rating 10A (2A)

8	○	SPARE
7	○	SPARE
6	○	L MAINS
5	○	N
4	○	N MAINS
3	○	COM
2	○	OFF
1	○	ON

Voltage-free switching unless 3–6 linked

Sangamo S250 series and S350 series

Electromechanical 24 hour time switch

3 TEMINAL			4 TERMINAL			
○	○	○	○	○	○	○
L	N	ON	COM	ON	L	N
					MAINS	

Link L–COM if required

On/off × 4 max.
H140 × W97 × D102
Switch rating 20A
Conduit box available, FD 930

Sangamo S408 Form 2

Electromechanical 24 hour time switch

	○	ON
	○	COM
N	○	
N	○	MAINS
L	○	MAINS

Voltage-free switching unless L–COM linked

On/off × pegs
H105 × W105 × D60
Switch rating 15A

Sangamo S408 Form 5

Electromechanical 24 hour time switch

○	ON
○	N
○	L

On/off × pegs
H105 × W105 × D60
Switch rating 15A

Sangamo S408 Form 6

Electromechanical 24 hour time switch

○	ON
○	COM
○	N MAINS
○	L MAINS

Voltage-free switching unless L–COM linked

On/off × pegs
H105 × W105 × D60
Switch rating 15A

Sangamo 409 FI

Electromechanical 24 hour Basic programmer

1	○	N	N	○	MAINS
2	○	HW ON	L	○	MAINS
3	○	ON	5	○	ROOM STAT OUT
4	○	CH ON	6	○	ROOM STAT IN

If no room stat link 5–6

On/off × 2
H106 × W160 × D67
Switch rating 10A (3A)

Sangamo 409 F3

Electromechanical 24 hour Full programmer

			1	○	CH ON
3	○	N	2	○	CH OFF
6	○	N MAINS	4	○	HW ON
7	○	L MAINS	5	○	HW OFF

On/off × 2
H106 × W160 × D67
Switch rating 10A (3A)

Sangamo 409 F4

Electromechanical 24 hour Basic programmer

1	○	N	N	○	MAINS
2	○	HW ON	L	○	MAINS
3	○	ON	5	○	ROOM STAT OUT
4	○	CH ON	6	○	ROOM STAT IN

If no room stat link 5–6

On/off × 2
H106 × W160 × D67
Switch rating 10A (3A)

Sangamo 409 F5

Electromechanical 24 hour Electricaire control

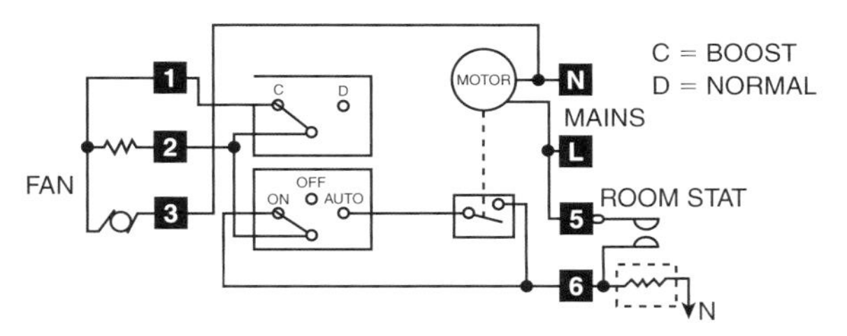

Figure 2.7

On/off × 2
H106 × W160 × D67
Switch rating 10A (3A)

Sangamo 409 F6

Electromechanical 24 hour warm air control

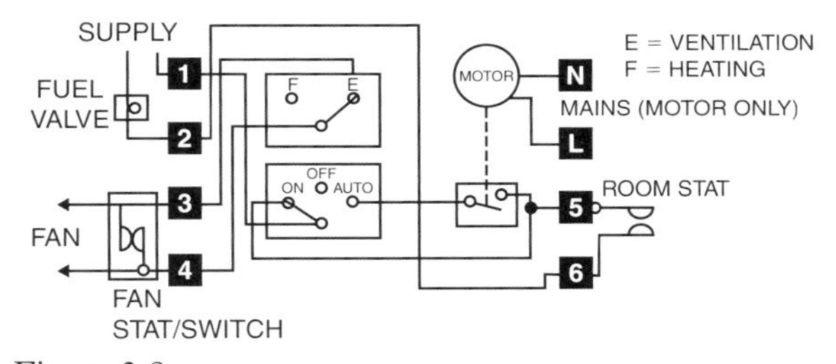

Figure 2.8

With provision for ventilation.

On/off × 2
H106 × W160 × D67
Switch rating 10A (3A)

Sangamo 409 F7

Electromechanical 24 hour priority programmer

For diagram see Figure 2.18, page 60

On/off × 2
H106 × W160 × D67
Switch rating 10A (3A)

Sangamo 409 F8

Electromechanical 24 hour time switch

1	○	SPARE	N	○	N MAINS
2	○	ON	L	○	L MAINS
3	○	N	5	○	ROOM STAT OUT
			6	○	ROOM STAT IN

If no room stat link 5–6

On/off × 2
H106 × W160 × D67
Switch rating 10A (3A)

Sangamo 410 F1

Electromechanical 24 hour Full programmer

8	○	CH ON
7	○	CH OFF
6	○	CH COM
5	○	N
4	○	N MAINS
3	○	L MAINS
2	○	HW OFF
1	○	HW ON

Link 3–6

On/off × 2
H85 × W138 × D46
Switch rating 10A (2A)

Sangamo 410 F2

Electromechanical 24 hour Basic programmer

8	○	CH ON
7	○	CH OFF
6	○	L MAINS
5	○	N MAINS
4	○	N
3	○	HW COM
2	○	HW OFF
1	○	HW ON

Link 3–6

On/off × 2
H85 × W138 × D46
Switch rating 10A (2A)

Sangamo 410 F3

Electromechanical 24 hour programmer

Specification as 410 F1 but labelled Zone 1 and Zone 2 instead to hot water and central heating

Sangamo 410 F4 (early model)

Electromechanical 24 hour Basic programmer

8	○	CH ON
7	○	
6	○	
5	○	N
4	○	N MAINS
3	○	L MAINS
2	○	
1	○	HW ON

Link 1–6

On/off × 2
H85 × W138 × D46
Switch rating 10A (2A)

Sangamo 410 F4

Electromechanical 24 hour Basic programmer

8	○	CH ON
7	○	CH OFF
6	○	CH COM
5	○	N
4	○	N MAINS
3	○	L MAINS
2	○	HW OFF
1	○	HW ON

Link 3–6

On/off × 2
H85 × W138 × D46
Switch rating 10A (2A)

Sangamo 410 F5

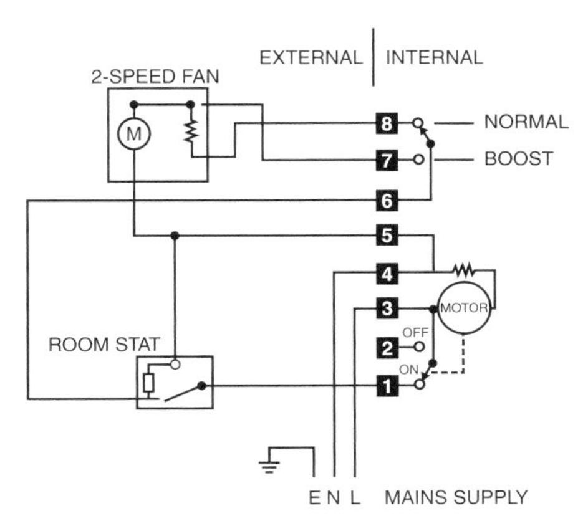

Figure 2.9

Electromechanical 24 hour Electricaire control

A single circuit programmer for controlling a two-speed fan on 'Electricaire' systems. It is fitted with an advance knob. The service knob allows a choice between 'Normal' or 'Boost' conditions.

On/off × 2
H85 × W138 × D46
Switch rating 10A (2A)

Sangamo 410 F6

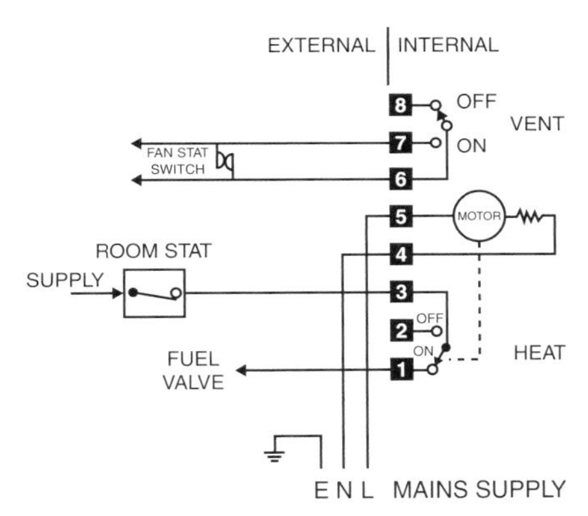

Figure 2.10

Electromechanical 24 hour warm air control

A single circuit programmer for controlling warm air systems with a provision for VENT (ventilation). It is fitted with an advance knob, programme selector knob and a manual selector knob to provide VENT 'ON' or 'OFF'.

On/off × 2
H85 × W138 × D46
Switch rating 10A (2A)

Sangamo 410 F7

For diagram see Figure 2.19, page 60

Electromechanical 24 hour priority programmer

On/off × 2
H85 × W138 × D46
Switch rating 10A (2A)

Sangamo 410 F8

8	○	SPARE
7	○	SPARE
6	○	SPARE
5	○	N
4	○	N MAINS
3	○	L MAINS
2	○	OFF
1	○	ON

Electromechanical 24 hour time switch

On/off × 2
H85 × W138 × D46
Switch rating 10A (2A)

Sangamo 410 F9

For diagram see Figure 2.25, page 62

Electromechanical 24 hour Full programmer

For fully pumped systems utilizing ACL/Tower Biflo mid-position valve.

On/off × 2
H85 × W138 × D46
Switch rating 10A (2A)

Sangamo 414 twin set

8	○	ZONE 1 ON
7	○	ZONE 1 OFF
6	○	L MAINS
5	○	N MAINS
4	○	N
3	○	ZONE 2 COM
2	○	ZONE 2 OFF
1	○	ZONE 2 ON

Link 3–6

Electromechanical 24 hour two-zone two-clock programmer

The programmer is fitted with two time switch dials on the left-hand side, each having two 'on' and two 'off' levers. The right-hand side is divided into two, the upper portion carrying the knobs and indicator associated with the upper time switch dial (Zone 1) and the lower portion with the lower dial (Zone 2).

On/off × 2 ea.
H85 × W138 × D61
Switch rating 10A (2A)

Sangamo 440

1	2	3	4	5	6	7	8
○	○	○	○	○	○	○	○
ON	SP	COM	N MAINS	SP	L MAINS	SP	SP

Voltage-free switching unless 3–6 linked

Electromechanical 24 hour time switch

On/off × 2
H85 × W138 × D46
Switch rating 10A (2A)

Sangamo 931091

N		N	L	1	2
○	—	○	○	○	○
		MAINS		ON	COM

Voltage-free switching unless L–2 linked

Electromechanical 24 hour time switch

On/off × multiple
H84 × W167 × D44
Switch rating 6A

Sangamo 931092

N		N	L	1	2
○	—	○	○	○	○
		MAINS		HW ON	CH ON

Electromechanical 24 hour Basic programmer

On/off × multiple
H84 × W167 × D44
Switch rating 6A

Sangamo 931093

N ○ —— N ○   L ○   1 ○   2 ○
MAINS   HW ON   CH ON

Electromechanical 24 hour Full programmer

On/off × multiple
H84 × W167 × D44
Switch rating 6A

Sangamo S610

3 TERMINAL: N, L, ON
4 TERMINAL: N, COM, L, ON

Link L–COM if required

Electromechanical 24 hour time switch

On/off × 4
H139 × W82 × D70
Switch rating 30A
Conduit box available – FD 1510

Sangamo S611

As S610 with day omittance device

Sangamo Set 1

As Randall Set 1

Sangamo Set 2

As Randall Set 2

Siemens RWB 27

Wiring as Landis & Staefa RWB30

Electronic 24 hour or 5/2 day or 7 day time switch

On/off × 3
H90 × W145 × D34
Switch rating 6A (2A)

Siemens RWB 29

Wiring as Landis & Staefa RWB40

Electronic 24 hour or 5/2 day or 7 day Basic/Full programmer

On/off × 3
H90 × W145 × D34
Switch rating 6A (2A)

SMC

For diagrams see Figures 2.16 and 2.17, page 59; also Figures 12.48 and 12.49, pages 254

The SMC programmers used in the SMC control pack (one boiler and two pumps, room and cylinder thermostats but no motorized valves) were manufactured by Horstmann. The first one was based on the 423 Pearl Auto and the next one was based on the 424 Emerald. They both contained a boiler switching relay. To replace, it is necessary to use either the new SMC wiring centre, incorporating a relay, or use a Basic or Full facility programmer and an external relay.

Smiths Centroller 10

Electromechanical 24 hour Basic programmer

Terminals 1 and 4 are internally linked and have no other connection

On/off × 2
H130 × W186 × D76
Switch rating 6A

Smiths Centroller 30

Electromechanical 24 hour time switch

1	2	3	4	5	6
N	L	SPARE	ON	ON	SPARE
MAINS					

Terminals 4 and 5 are internally linked

On/off × 2
H98 × W156 × D72
Switch rating 6A

Smiths Centroller 30+

Electromechanical 24 hour priority programmer

For diagram see Figure 2.20, page 60

On/off × 2
H98 × W156 × D72
Switch rating 6A

Smiths Centroller 40

Electromechanical 24 hour Basic programmer

1	2	3	4	5
N	L	SPARE	HW ON	CH ON
MAINS				

On/off × 2
H105 × W152 × D80
Switch rating 6A

Smiths Centroller 40+

Electromechanical 24 hour priority programmer

For diagram see Figure 2.21, page 61

On/off × 2
H98 × W156 × D72
Switch rating 6A

Smiths Centroller 50

Electromechanical 24 hour time switch

1	2	3	4	5
N	L	COM	ON	N
MAINS				

Voltage-free switching unless 2–3 linked

On/off × 2
H105 × W152 × D72
Switch rating 6A

Smiths Centroller 60

Electromechanical 24 hour Basic programmer

1	2	3	4	5
○	○	○	○	○
N	L	SPARE	HW	CH
MAINS			ON	ON

On/off × 2
H105 × W152 × D80
Switch rating 6A

Smiths Centroller 70

Electromechanical 24 hour Basic programmer

1	2	3	4	5	6
○	○	○	○	○	○
N	L	SPARE	CH	HW	SPARE
MAINS			ON	ON	

On/off × 2
H98 × W156 × D72
Switch rating 6A

Smiths Centroller 90

Electromechanical 24 hour Full programmer

1	2	3	4	5	6
○	○	○	○	○	○
N	L	SPARE	CH	HW	SPARE
MAINS			ON	ON	

On/off × 2
H98 × W156 × D72
Switch rating 6A

Smiths Centroller 100

Electromechanical 24 hour Basic programmer

N	N	L	1	2	3	4
○	○	○	○	○	○	○
	N	L	SPARE	CH	HW	SPARE
	MAINS			ON	ON	

On/off × 2
H130 × W186 × D76
Switch rating 6A

Smiths Centroller 1000

Electronic 24 hour Basic/Full programmer

N	L	1	2	3	4
○	○	○	○	○	○
MAINS		HW	CH	HW	CH
		OFF	OFF	ON	ON

On/off × 2
H84 × W140 × D40
Switch rating 2A

Smiths Centroller 2000
3000

See **Programmers and time switches with inbuilt or external sensors or thermostats** (Chapter 3)

Smiths Supply Master CHP11

Electronic 24 hour Basic programmer

Incorporated in a switched fused spur

6	5	4	3	2	1
○	○	○	○	○	○
L	N	HW	N	SPARE	CH
MAINS		ON			ON

On/off × 4
H85 × W85 × D37
Switch rating 3A

Smiths Supply Master CHP17

7 day version of CHP11

Smiths Supply Master FST11

Electronic 24 hour time switch

On/off × 4
H85 × W85 × D37
Switch rating 13A

Incorporated in a switched fused spur

6	5	4	3	2	1
○	○	○	○	○	○
L	N	N	L	COM	ON
MAINS					

Voltage-free switching unless 3–2 linked

Smiths Supply Master FST17

7 day version of FST11

Sopac 24 Hour Fuelminder

Electromechanical 24 hour Basic programmer

On/off × 48
H159 × W83 × D42
Switch rating 15A (2A)

1	○	CH ON
2	○	N MAINS
3	○	L MAINS
4	○	HW ON
5	○	DO NOT USE

Sopac 7D Fuelminder

As 24 hour but with 7 day setting facility with on 6 on/offs per day

Southern Digital

A range of time switches and controls not specifically designed for central heating but larger loads, e.g. immersion heaters

SUGG Supaheat

Electromechanical 24 hour Full programmer

On/off × 2

1	2	3	4	5	6	7
○	○	○	○	○	○	○
CH ON	CH OFF	N	HW OFF	HW ON	N	L
					MAINS	

Sunvic CB2201

Electromechanical 24 hour Full programmer

This unit is only available as a Clockbox 2 spare

Sunvic DHP 1201 Libra

Electromechanical 24 hour Basic programmer

On/off × 6
H125 × W190 × D64
Switch rating 3A

1	2	3	4	5	6	7	8
○	○	○	○	○	○	○	○
N	L	CH ON	COM	CH OFF	HW ON	COM	HW OFF
MAINS							

Voltage-free switching unless 2–4–7 linked

Sunvic DHP 2201

Electronic 24 hour Full programmer

1	2	3	4	5	6	7	8
N	L	CH ON	COM	CH OFF	HW ON	COM	HW OFF
MAINS							

Voltage-free switching unless 2–4–7 linked

On/off × 6
H125 × W190 × D64
Switch rating 3A

Sunvic ET 1401/ET 1402

Electronic 24 hour Full programmer

1	2	3	4	5	6	7	8
N	L	COM	CH ON	CH OFF	COM	HW ON	HW OFF
MAINS							

Voltage-free switching unless 2–3–6 linked

On/off × 4
H110 × W180 × D65
Switch rating 5A (1A)

Sunvic ET 1451

As ET 1401 with battery reserve and advance features

Sunvic MP2

Electromechanical 24 hour Full programmer

			L	L	N	N
HW OFF	HW ON	CH ON	MAINS			

On/off × 2
H100 × W202 × D56
Switch rating 5A (1A)

Sunvic Select 107

Electronic 24 hour or 5/2 day or 7 day time switch

N	L	1	2	3	4
MAINS		COM	OFF	ON	SPARE

Voltage-free switching unless L–1 linked

On/off × 2
H82 × W135 × D36
Switch rating 3A (1A)

Sunvic Select 207

Electronic 24 hour or 5/2 day or 7 day Basic/Full programmer

N	L	1	2	3	4
N	L	HW OFF	CH OFF	HW ON	CH ON
MAINS					

On/off × 2
H82 × W135 × D36
Switch rating 3A (1A)

Sunvic SP 20

Electronic 24 hour time switch

L	N	3	4	5
MAINS		COM	SPARE	ON

Voltage-free switching unless L–3 linked

On/off × 2
H91 × W161 × D42
Switch rating 5A (1A)

Sunvic SP 25

Electronic 24 hour Basic programmer

1	2	L	N	E	S	S	3	4	5
HW OFF	HW ON		MAINS		SPARE		NOT USED		CH ON

On/off × 2
H91 × W161 × D42
Switch rating 5A (1A)

Sunvic SP 30

Electronic 24 hour Basic programmer

1	2	L	N	E	S	S	3	4	5
HW OFF	HW ON		MAINS		SPARE		NOT USED		CH ON

This unit is similar to the SP 25 with extra features which may be useful for the elderly or visually handicapped, including setting tones.

On/off × 2
H91 × W161 × D42
Switch rating 5A (1A)

Sunvic SP 35

Electronic 24 hour time switch

L	N	3	4	5
MAINS		COM	SPARE	ON

Voltage-free switching unless L–3 linked

On/off × 3
H91 × W161 × D42
Switch rating 6A (1A)

Sunvic SP 50

Electronic 24 hour Basic/Full programmer

1	2	L	N	E	S	S	3	4	5
HW OFF	HW ON	L	N MAINS	E	SPARE		CH COM	CH OFF	CH ON

Link L–3

Facility for setting hot water and heating at different times weekly in Full mode.

On/off × 2
H91 × W161 × D42
Switch rating 6A (1A)
For Basic control cut link at rear of programmer

Sunvic SP 100

Electronic 7 day Basic/Full programmer

1	2	L	N	E	S	S	3	4	5
HW OFF	HW ON	L	N MAINS	E	SPARE		CH COM	CH OFF	CH ON

Link L–3

Facility for setting hot water and heating at different times to each other every day in Full mode

On/off × 2
H91 × W161 × D42
Switch rating 6A (1A)
For Basic control cut link at rear of programmer

Superswitch 1511

As Sopac Fuelminder 24H

Superswitch 1517

As Sopac Fuelminder 7D

Superswitch 1647

L	N	N	L1	L2
○	○	○	○	○
MAINS		LOAD	ON	SPARE

Electronic 24 hour time switch

On/off × 6
H90 × W162 × D75
Switch rating 15A (2)A

Superswitch 1657

L	N	N	L1	L2
○	○	○	○	○
MAINS		LOAD	HW ON	CH ON

Electronic 24 hour time switch

On/off × 6
H90 × W162 × D75
Switch rating 15A (2)A

Switchmaster 300

N	L	1	2	3	4
○	○	○	○	○	○
MAINS		ON	N	SPARE	COM

Links were fitted N–2 and L–4 but may have been removed depending on use. On the later ACL/Switchmaster version no links were fitted.
Voltage-free switch

Electromechanical 24 hour time switch

On/off × 2
H100 × W168 × D48
Switch rating 5A (2A)

Switchmaster 320

N	L	1	2	3	4
○	○	○	○	○	○
MAINS		CH ON	N	HW ON	L

Links were fitted N–2 and L–4 on backplate

Electromechanical 24 hour Basic programmer

On/off × 2
H100 × W168 × D48
Switch rating 5A (2A)

Switchmaster 350

N	L	1	2	3	4
○	○	○	○	○	○
MAINS		CH ON	N	HW ON	L

Link were fitted N–2 and L–4 on backplate

Electromechanical 24 hour Basic programmer

On/off × 2
H100 × W168 × D48
Switch rating 5A (2A)

Switchmaster 400

N	L	1	2	3	4
○	○	○	○	○	○
MAINS		CH ON	SPARE	HW ON	HW OFF

Electromechanical 24 hour Basic programmer

On/off × 2
H100 × W168 × D48
Switch rating 5A (2A)

Switchmaster 500

For diagram see Figure 2.22, page 61

Electromechanical 24 hour priority programmer

On/off × 2
H100 × W168 × D48
Switch rating 5A (2A)

Switchmaster 600

N	L	1	2	3	4
MAINS		CH ON	SPARE	HW ON	SPARE

Electromechanical 24 hour Basic programmer

On/off × 2
H100 × W168 × D48
Switch rating 5A (2A)

Switchmaster 800

As 805 with neon indicators

Switchmaster 805

N	L	1	2	3	4
MAINS		CH ON	CH OFF	HW ON	HW OFF

Electromechanical 24 hour Full programmer

On/off × 2
H100 × W168 × D48
Switch rating 5A (2A)

Switchmaster 900

N	L	1	2	A	B	C	3	4
MAINS		CH ON	CH OFF		SPARE		HW ON	HW OFF

Electromechanical 24 hour Basic/Full programmer

On/off × 2
H138 × W83 × D53
Switch rating 5A (2A)
To adjust from Basic to Full mode, turn screw at rear of programmer from 10 to 16

Switchmaster 905

As 900 with different styling

Switchmaster 950

N	L	1	2	A	B	C
MAINS		ON	OFF		SPARE	

Electromechanical 24 hour time switch

On/off × 2
H138 × W83 × D53
Switch rating 5A (2A)

Switchmaster 980 Combi

N	L	1	2	A	B	C	3	4
MAINS		ON	CH		SPARE			COM

Voltage-free switching unless L–4 linked

Electromechanical 24 hour time switch

On/off × 2
H138 × W83 × D53
Switch rating 5A (2A)

Switchmaster 3000

As 300 with Homewarm motif

Switchmaster 9000

N	L	1	2	A	B	C	3	4
MAINS		CH ON	CH OFF		SPARE		HW ON	HW OFF

Electronic 24 hour Basic/Full programmer

On/off × 2
H136 × W83 × D454
Switch rating 3A (2A)
To adjust from Basic to Full mode, move the slider at the rear of the programmer to the left.

Switchmaster 9001

As 9000 with different styling

Switchmaster Sonata

L	N	1	2	4	6
MAINS		HW ON	HW OFF	CH ON	CH OFF

Electronic 7 day Full programmer

On/off × 3
H95 × W161 × D40
Switch rating 3A (2A)

Thorn

A version of the Randall 4033 was marketed with a Thorn cover and mounted horizontally

Thorn Microtimer

As Honeywell ST 699 B

Tower DP 72

N	L	1	2	3	4
MAINS		HW OFF	CH OFF	HW ON	CH ON

Facility for setting hot water and heating at different times for each other every day in Full mode

Electronic 7 day Basic/Full programmer

On/off × 2
H79 × W163 × D50
Switch rating 5A (2A)
For Full mode remove tab switch at rear of programmer

Tower DT 71

N	L	1	2	3	4
MAINS		SPARE	OFF	COM	ON

Voltage-free switching unless L–3 linked

Electronic 7 day time switch

On/off × 2
H79 × W163 × D50
Switch rating 5A (2A)

Tower FP

As ACL FP

Tower MP

As ACL MP

Tower QE1

Electronic 7 day time switch

N	L	1	2	3	4
○	○	○	○	○	○
MAINS		SPARE	OFF	COM	ON

Voltage-free switching unless L–3 linked

On/off × 2
H85 × W155 × D45
Switch rating 5A (2A)

Tower QE2

Electronic 7 day Basic/Full programmer

N	L	1	2	3	4
○	○	○	○	○	○
MAINS		HW OFF	CH OFF	HW ON	CH ON

On/off × 2
H85 × W155 × D45
Switch rating 5A (2A)
To change from Basic to Full move slider at rear of programmer

Tower TC

As ACL TC

Tower T2000

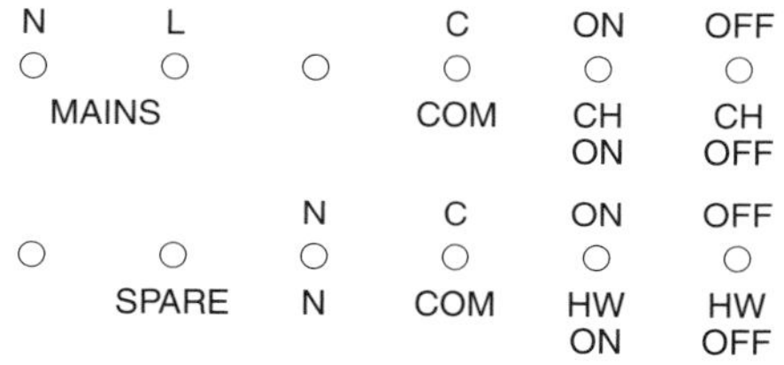

Electronic 24 hour Full programmer

With facility for 5/2 day setting.

N	L		C	ON	OFF
○	○	○	○	○	○
MAINS			COM	CH ON	CH OFF
		N	C	ON	OFF
○	○	○	○	○	○
	SPARE	N	COM	HW ON	HW OFF

Voltage-free switching unless L–C–C linked

On/off × 2
H75 × W156 × D50
Switch rating 6A (2A)

Tower T2001

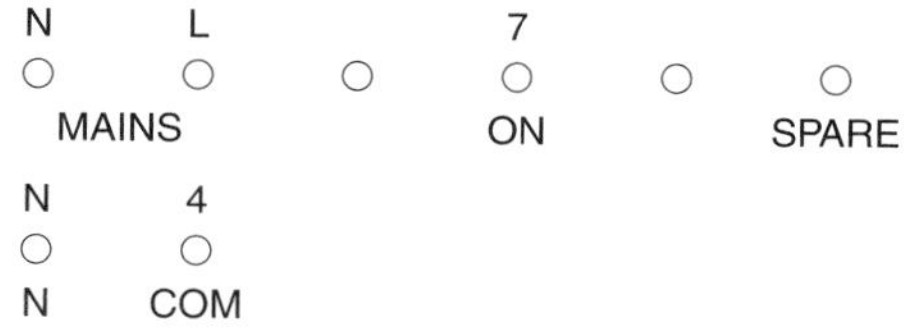

Electromechanical 24 hour time switch

N	L		7		
○	○	○	○	○	○
MAINS			ON		SPARE
N	4				
○	○				
N	COM				

Voltage-free switching unless L–4 linked

On/off × 4
H80 × W165 × D55
Switch rating 6A (2A)

Tower T2002

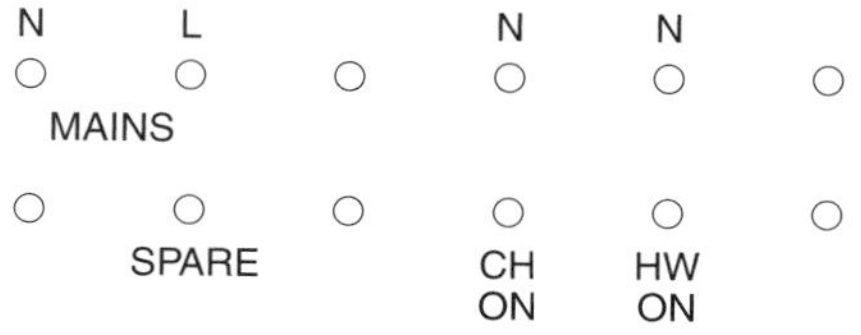

Electromechanical 24 hour Basic programmer

N	L		N	N	
○	○	○	○	○	○
MAINS					
○	○	○	○	○	○
	SPARE		CH ON	HW ON	

On/off × 4
H80 × W165 × D55
Switch rating 6A (2A)

Tower T2003

N	L		c	ON	OFF
○	○	○	○	○	○
MAINS			COM	CH ON	CH OFF
		N	c	ON	OFF
○	○	○	○	○	○
SPARE		N	COM	HW ON	HW OFF

Voltage-free switching unless L–C–C linked

Electromechanical 24 hour Basic/Full programmer

On/off × 4
H80 × W165 × D55
Switch rating 6A (2A)
For Full mode remove plastic switch slide covers

Trac TP 10

As Flash 31031

Trac TP 20

As Flash 31032

Trac TP 20/7

As Flash 31732

Trac TP 30

As Flash 31033

Trac TP 30E

As Switchmaster 9001

Venner CHC/2

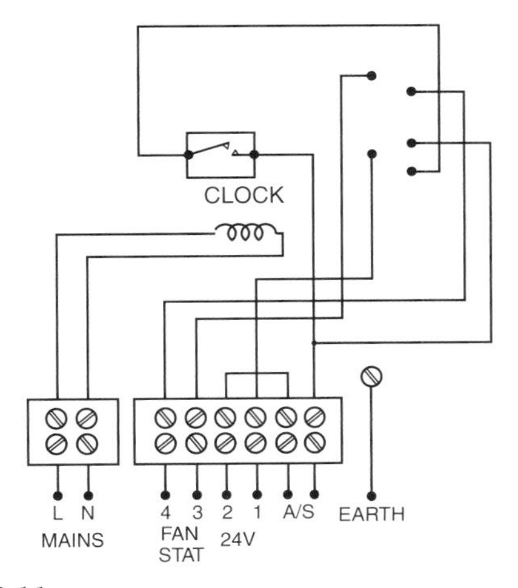

Figure 2.11

Electromechanical 24 hour warm air control

On/off × 2
H84 × W135 × D88
Switch rating 5A (2A)

Venner CHC/GW

For diagram see Figure 2.23, page 61

Electromechanical 24 hour priority programmer

On/off × 2
H84 × W135 × D88
Switch rating 5A (2A)

Venner CHC/W2

Electromechanical 24 hour Basic programmer

As		4	3	2	1	N	L
ROOM STAT		N	CH ON	N	HW ON	MAINS	

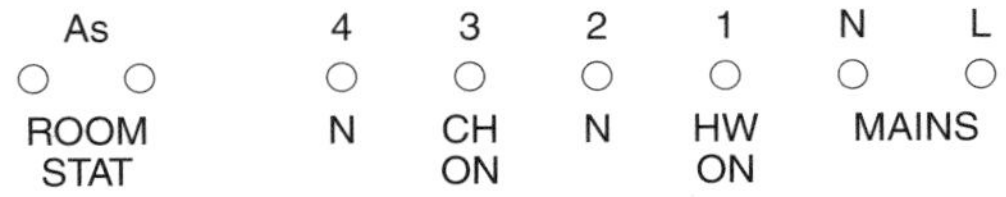

On/off × 2
H84 × W135 × D88
Switch rating 5A (2A)

Venner T10

Electromechanical 24 hour time switch

	3	2	1		N	L1	L	E
	ON	COM	OFF		MAINS		MAINS	

Voltage-free switching unless L1–2 linked

On/off × 2
H162 × W100 × D63
Switch rating 15A

Venner T20

Electromechanical 24 hour Basic programmer

6	5	4	3	2	1	E	N	L
ROOM STAT		CH ON	N	N	HW ON	E	MAINS	

Link 5–6 if no room stat

On/off × 2
H162 × W100 × D63
Switch rating 6A

Venner T30

Electromechanical 24 hour Full programmer

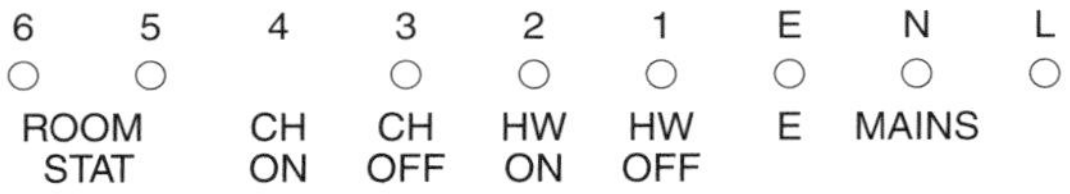

6	5	4	3	2	1	E	N	L
ROOM STAT		CH ON	CH OFF	HW ON	HW OFF	E	MAINS	

Link 5–6 if no room stat

On/off × 2
H162 × W100 × D63
Switch rating 6A

Venner Vennerette

Electromechanical 24 hour time switch

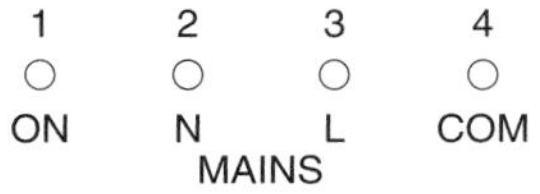

1	2	3	4
ON	N MAINS	L	COM

Voltage-free switching unless 3–4 linked

On/off × 2
H102 × W87 × D76
Switch rating 30A

Venner Venomise

Electromechanical 24 hour time switch

Terminal	Function
3	ON
2	COM
1	OFF
N	MAINS
L	MAINS

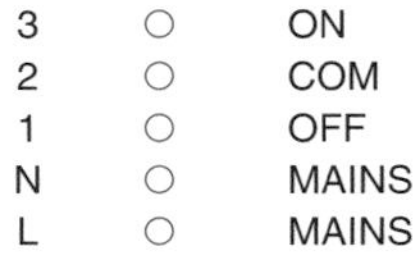

The Venomise is designed for control of an immersion heater or similar.

On/off × 2
H113 × W109 × D50
Switch rating 16A

Venner Venneron

Electromechanical 24 hour time switch

1	2	3	4
○	○	○	○
COM	L	N	ON
	MAINS		

Voltage-free switching unless 1–2 linked

On/off × 2
H115 × W88 × D82
Switch rating 15A

Venner Venotime

Electromechanical 24 hour time switch

3	○	ON
2	○	COM
1	○	OFF
N	○	MAINS
L	○	MAINS

Voltage-free switching unless L–2 linked

Venotime Selective has day omission device.

On/off × 2
H113 × W109 × D50
Switch rating 16A

Venner Venotrol

Electromechanical 24 hour Basic programmer

On/off × 2
H200 × W118 × D80

Venner Venotrol 80

Electromechanical 24 hour Basic programmer

A/S		5	4	3	2	1	N	L
○	○		○	○	○	○	○	○
ROOM STAT		CH ON	N	N	HW ON	N	MAINS	

Fit link if no room stat

On/off × 2
H105 × W170 × D50

Venner Venotrol 80M

Electromechanical 24 hour Full programmer

A/S		5	4	3	2	1	N	L
○	○		○	○	○	○	○	○
ROOM STAT		CH ON	CH OFF	N	HW ON	HW OFF	MAINS	

Fit link if no room stat

On/off × 2
H105 × W170 × D50

Venner Venotrol 80P

Electromechanical 24 hour Full programmer

A/S		5	4	3	2	1	N	L
○	○		○	○	○	○	○	○
ROOM STAT		CH ON	CH OFF	N	HW ON	HW OFF	MAINS	

Fit link if no room stat

On/off × 2
H105 × W170 × D50

Venner Venotrol 80PM

Electromechanical 24 hour Full programmer

A/S		5	4	3	2	1	N	L
○	○		○	○	○	○	○	○
ROOM STAT		CH ON	CH OFF	N	HW ON	HW OFF	MAINS	

Fit link if no room stat

On/off × 2
H105 × W170 × D50

Venner Venotrol 90

Electromechanical 24 hour Basic programmer

A/S		5	4	3	2	1	N	L
○	○	○	○	○	○	○	○	○
ROOM STAT		CH ON	CH OFF	N	HW ON	N	MAINS	

Fit link if no room stat

On/off × 2
H105 × W170 × D50

Priority system programmers

Replacing a priority programmer with a double-circuit programmer

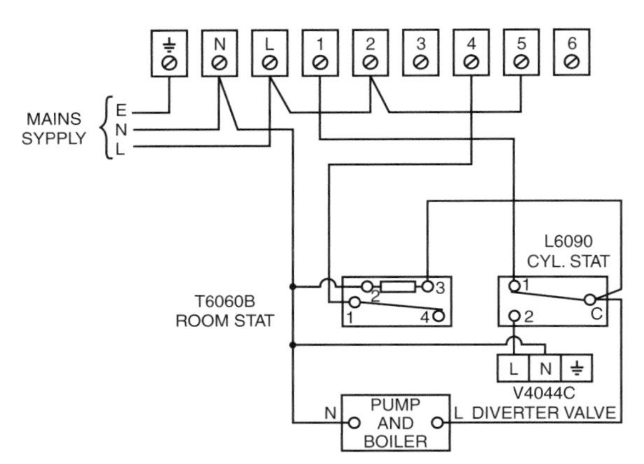

Figure 2.12 *Horstmann 425/525 with Honeywell motorized valve and thermostats*

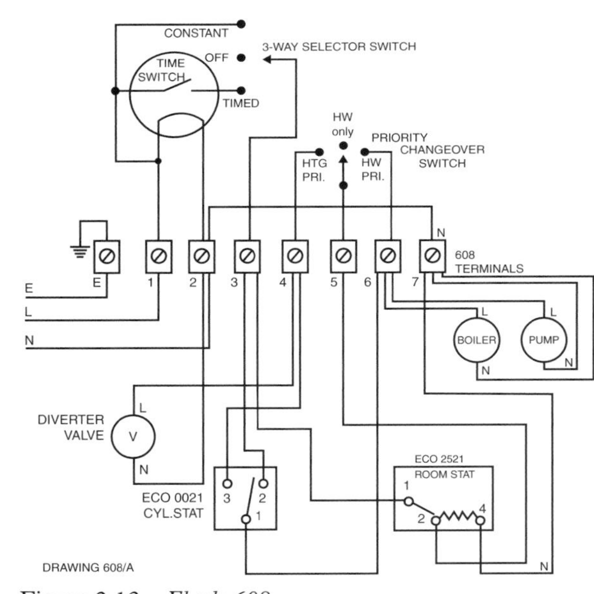

Figure 2.13 *Eberle 608*

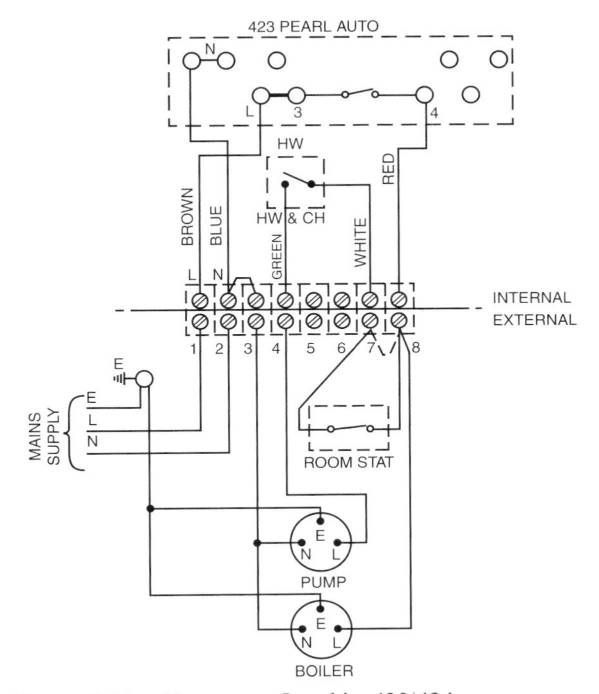

Figure 2.14 *Horstmann Sapphire 423/424*

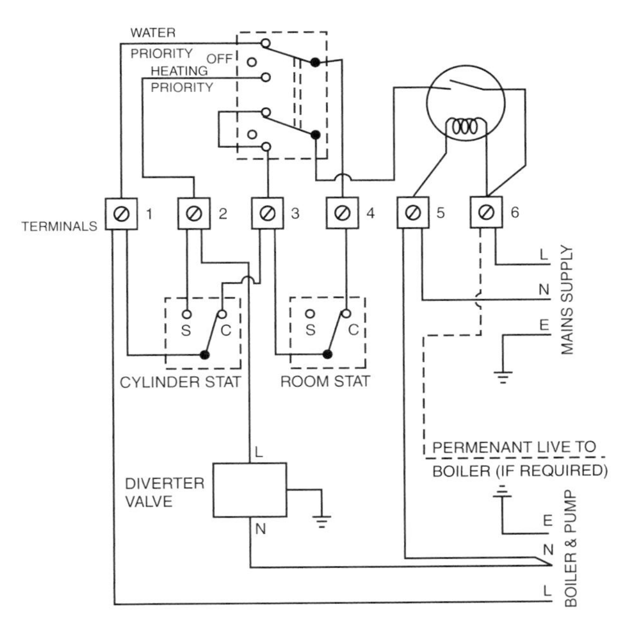

Figure 2.15 *Randall 106*

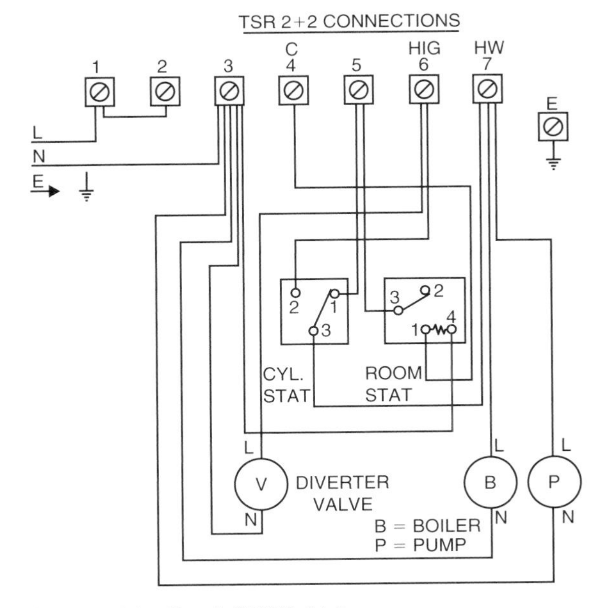

Figure 2.16 *Randall TSR 2+2*

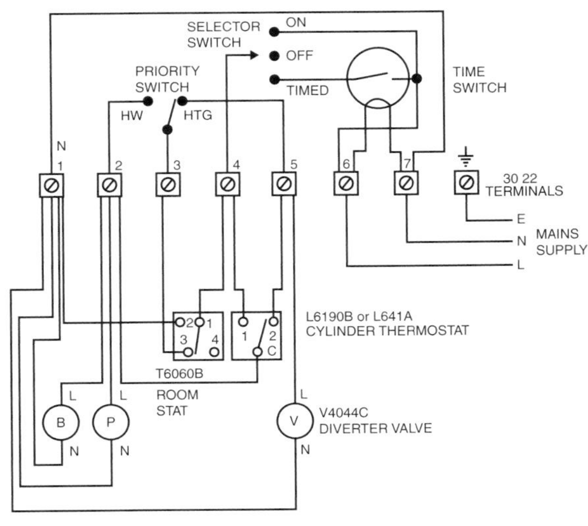

Figure 2.17 *Randell 3022*

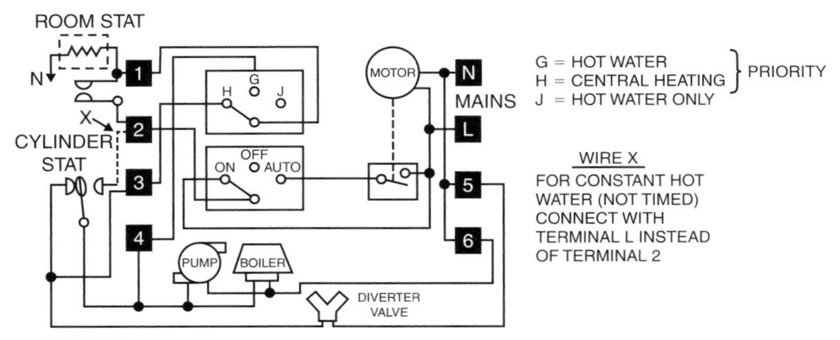

Figure 2.18 *Sangamo 409 F7*

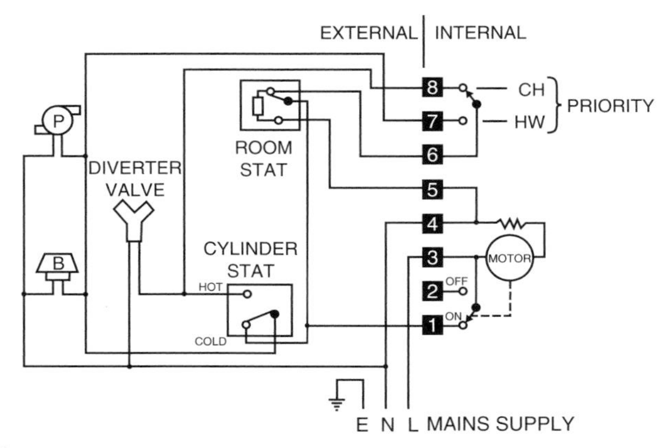

Figure 2.19 *Sangamo 410 F7*

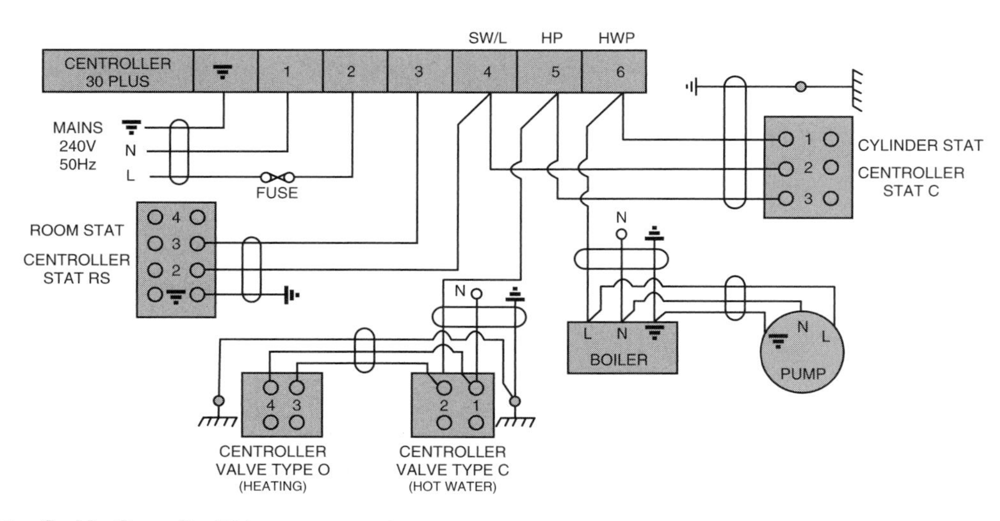

Figure 2.20 *Smiths Centroller 30+*

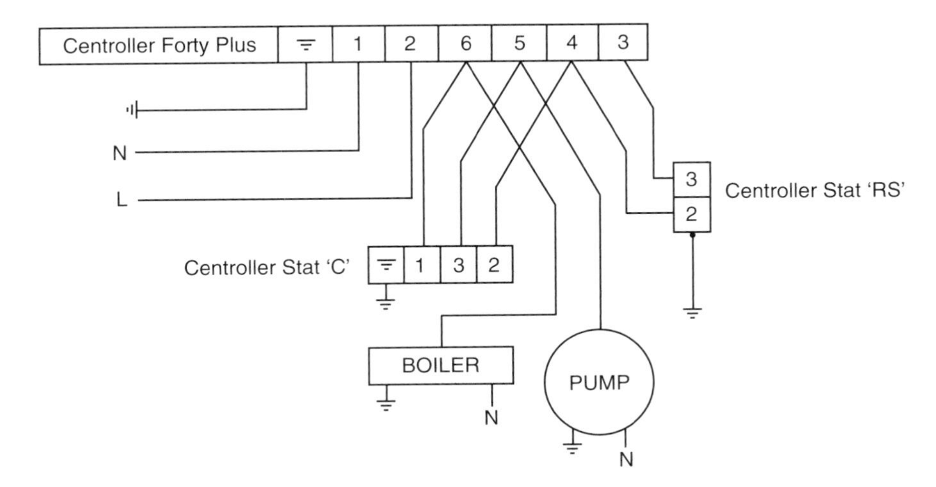

Figure 2.21 *Smiths Centroller 40+*

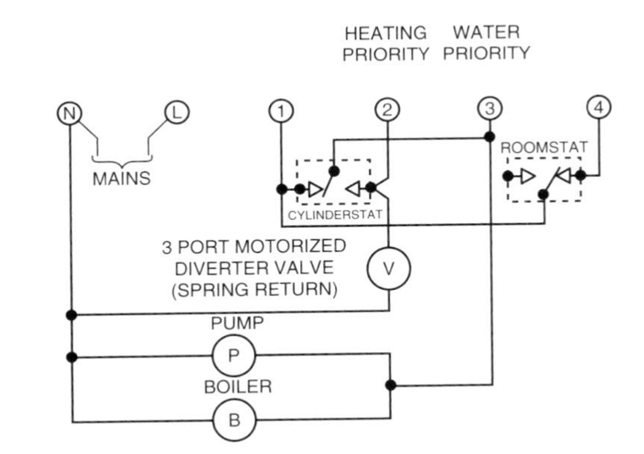

Figure 2.22 *Switchmaster 500*

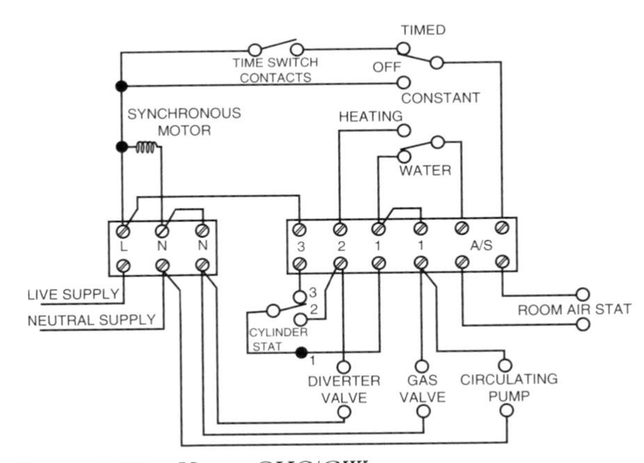

Figure 2.23 *Venner CHC/GW*

Programmers specifically designed for used with the ACL 672 BR 340 Biflo motorized valve

Others were suitable for this and other systems. See motorized valves, Chapter 6.

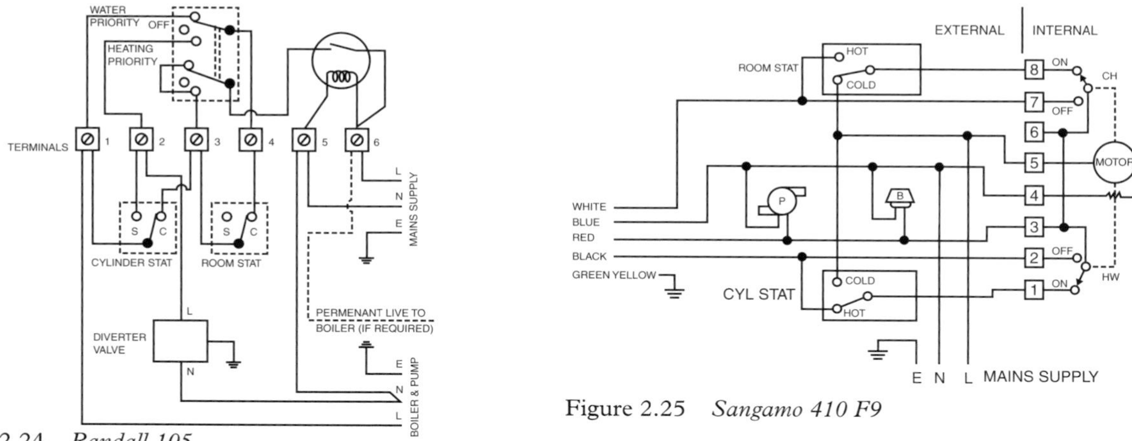

Figure 2.24 *Randall 105*

Figure 2.25 *Sangamo 410 F9*

SMC Control pack wiring diagrams utilizing Horstmann programmers

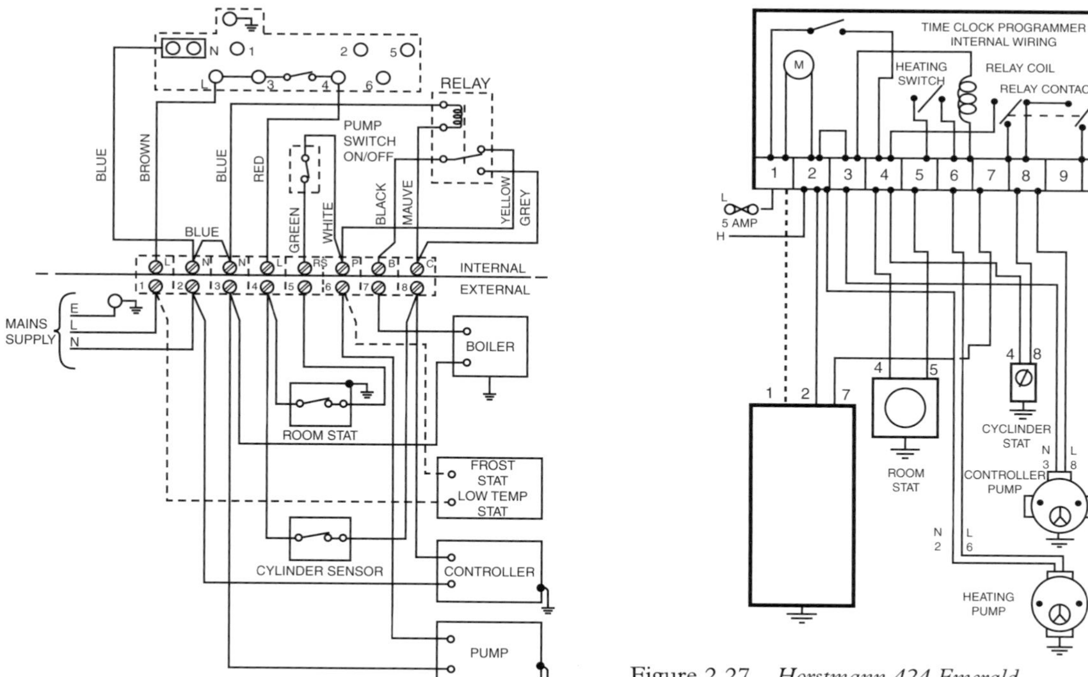

Figure 2.26 *Horstmann 423 Pearl Auto*

Figure 2.27 *Horstmann 424 Emerald*

3

Programmers and time switches with inbuilt or external sensors or thermostats

ACL CC 240
CC 520

Electronic comfort controller
(full control only)

The CC 240/520 allows switching times and room temperature to be selected from a central point. A remote sensor DR 174 is used to detect and control room temperature. The hot water temperature can be controlled using a conventional cylinder thermostat. The CC 240 has 24 hour programming and the CC 520 has weekday/weekend programming.

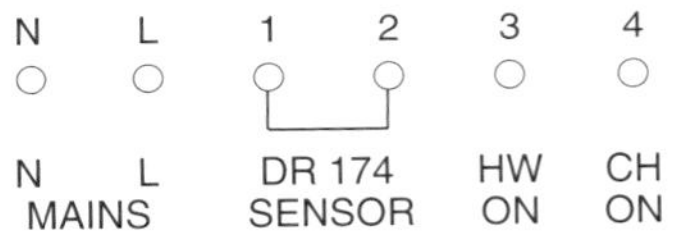

N	L	1	2	3	4
○	○	○	○	○	○
N	L	DR 174		HW	CH
MAINS		SENSOR		ON	ON

Terminals 4 and 5 are internally linked

Scale 5–30°C
H93 × W148 × D31
Switch rating 2A (1A)
On/offs CC 240 × 2
CC 520 × 3

ACL CT 171
CT 172

Programmable electronic thermostats with low-temperature protection

The CT 171/172 allow switching times and room temperature to be selected from a single point. An inbuilt sensor is used to detect and control room temperature. In the OFF position central heating will remain off unless the temperature drops to the low-temperature point, in which case it will automatically switch on again.

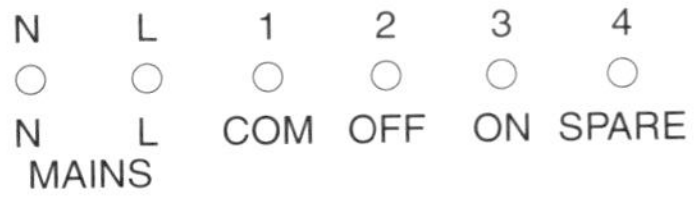

N	L	1	2	3	4
○	○	○	○	○	○
N	L	COM	OFF	ON	SPARE
MAINS					

Voltage-free switching unless L–1 linked

Scale CT 171 5–30°C
CT 172 16–28°C
Low-temperature protection
CT 171 7°C
CT 172 16°C
H87 × W170 × D74
Switch rating 2A (1A)
On/offs × 2
Batteries required 3 × AAA

ACL CT 174

Electronic clock thermostat

The CT 174 allows switching times and room temperature to be selected from a central point. A remote sensor DR 174 is used to detect and control room temperature. The CT 174 has 7 day programming.

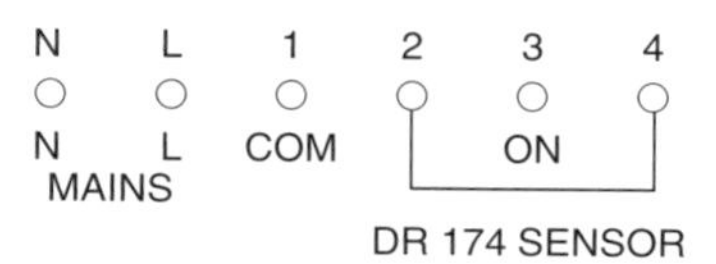

Voltage-free switching unless L–1 linked

Scale 16–28°C (scaled 1–5)
H87 × W170 × D47
Switch rating 2A (1A)
On/offs × 2
Batteries required 3 × AAA

ACL ILP 112

Electronic 24 hour Basic/Full programmer
(with optimum start feature through inbuilt sensor)

N L 1 2 3 4
MAINS HW OFF CH OFF HW ON CH ON

On/offs × 2
H93 × W148 × D31
Switch rating 2A (1A)
Move slider at rear of programmer to G for Basic control or P for Full control.

ACL OCC 520

5/2 day version of OCC 720

ACL OCC 720

Electronic 7 day Full comfort controller
(with optimum start feature)

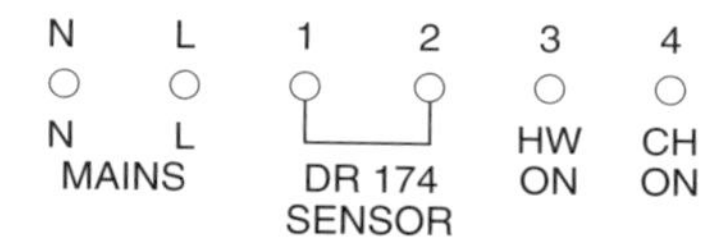

On/offs × 3
H93 × W148 × D31
Switch rating 2A (1A)

ACL OLP 552

Electronic weekday/weekend Basic/Full programmer (with optimum start feature)

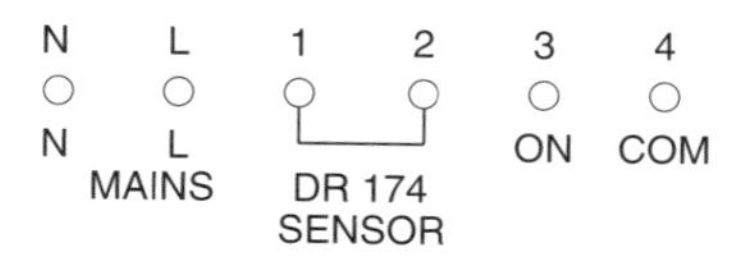

On/offs × 2
H93 × W148 × D31
Switch rating 2A (1A)
Move slider at rear of programmer to G for Basic control and P for Full control.

ACL OLP 722

7 day version of OLP 522

ACL OPT 170

Electronic clock thermostat
(with optimum start feature)

The OPT 170 allows switching times and room temperatures to be selected from a central point. A remote sensor DR 174 is used to detect and control room temperature. The optimum start feature automatically reduces warm up time for the heating system as weather becomes milder.

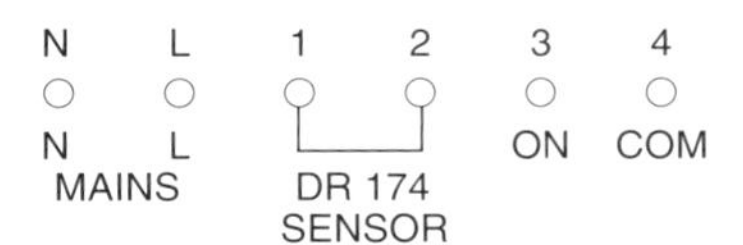

Voltage-free switching unless L–4 linked

Scale 5–30°C
H93 × W148 × D31
Switch rating 2A (1A)
On/offs × 2

ACL PT 110 PT 170

Programmable electronic thermostat

The PT 110/170 allows switching times and room temperature to be selected from a central point. A remote sensor DR 174 is used to detect and control room temperature. The PT 110 has a 24 hour programme and the PT 170 has a 7 day programme.

N L 1 2 3 4
N L
MAINS
DR 174
SENSOR
ON COM

Voltage-free switching unless L–4 linked

Scale 5–30°C
H93 × W148 × D31
Switch rating 2A (1A)
On/offs × 2

ACL PT 271 PT 371

Programmable electronic thermostat

The PT 271/371 allows switching times and room temperatures to be selected from a central point. An inbuilt sensor is used to detect and control room temperature. PT 271 programmer options are off/timed/high/low. When set to timed the programmer will turn on or off at the pre-programmed times and temperatures. There are four temperature setting periods. The PT 371 programme options are low/timed/medium/high. There are six temperatures setting periods.

N L 1 2 3 4
N L COM OFF ON SPARE
MAIN

Voltage-free switching unless L–1 linked

Scale 5–30°C
H93 × W148 × D31
Switch rating 2A (1A)

ACL-Drayton Digistat 2

24 hour programmable room thermostat

1 2 3
COM SAT DEM

Scale 5–30°C
H87 × W87 × D33
Switch rating 2A (1A)
On/offs × 2
Batteries required 2 × AA

ACL-Drayton Digistat 3

24 hour programmable room thermostat

7 day version of Digistat 2.

Danfoss-Randall HC75 HC75A

Electronic programmable room thermostat

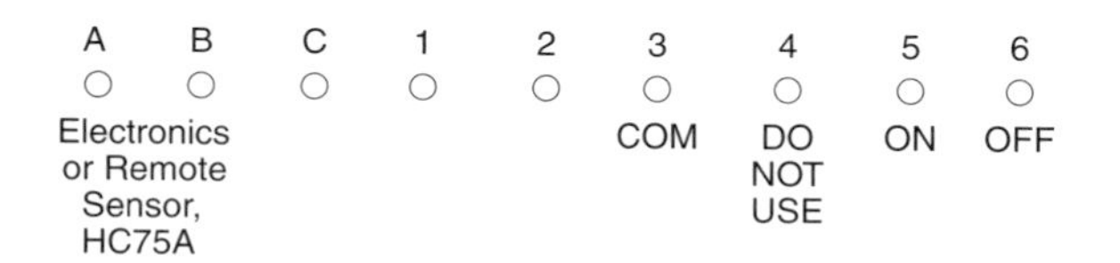

Scale 5–30°C
H88 × W135 × D32
Switch rating 2A (1A)
Batteries required 2 × AA
6 setting periods in 24 hours

Danfoss-Randall RT52

Set-back room thermostat with built in timer

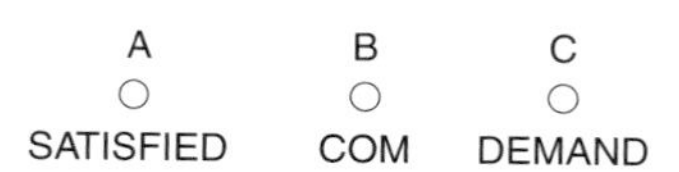

Scale 5–30°C
H88 × W110 × D28
Switch rating 6A (2A)
On/off × 2

Danfoss-Randall RT52-RF

Set-back thermostat with built-in timer and remote sensor

Receiver RX2

N	L	1	2	3	4
N	L	ZONE 2 ON	COM	ZONE 1 ON	ZONE 1 OFF
MAINS					

Scale 5–30°C
H88 × W138 × D32
Switch rating 3A (1A)
On/off × 2

Danfoss-Randall TP5E

Electronic programmable room thermostat

3	2	1
ON	OFF	COM

As TP5 with enhanced setting features

Scale 5–30°C
H81 × W99 × D34
Switch rating 6A (2A)
Up to 6 temperature changes per day
Batteries required 2 × AA

Danfoss-Randall TP5E-RF

Radio controlled programmable room thermostat

The TP5E-RF is a control for use in situations where it is not practical or acceptable to run cables between the room thermostat and the other control components in the system.

Using radio transmissions the TP5E-RF communicates instructions to a receiver unit, type RX, which can be located up to 30m away from the thermostat.

The receiver unit is available in three versions offering one, two or three zones of control, each zone requiring its own TP5E-RF thermostat. Any TP5E-RF can be tuned or re-tuned to any receiver channel.

The TP5E-RF thermostat offers identical functionality to that of the hard wired TP5E version, except that the relay function is moved from the thermostat to the RX receiver unit.

RX1

Scale 5–30°C
TP5E-RF H81 × W98 × D34
RX receiver H88 × W138 × D32
Switch rating 3A (1A)
Batteries required 2 × AA

RX2

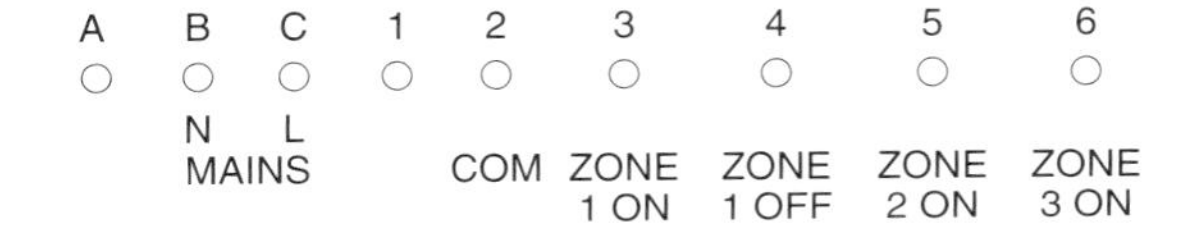

RX3

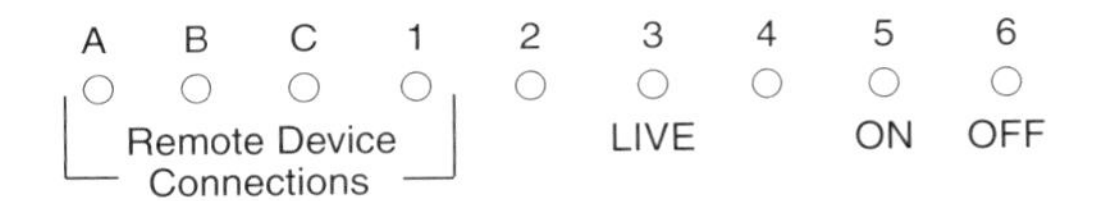

Danfoss-Randall TP75

7 day programmable room thermostat

Scale 5–30°C
H88 × W135 × D32
Switch rating 2A (1A)
Up to 6 temperature changes per day
Batteries required 2 × AA

Danfoss-Randall TP75M

7 day programmable room thermostat

E N L 1 2 3 4 5 6
MAINS COM OFF SP ON Remote Sensor

7 day version of TP75

Danfoss-Randall TP75H, TP75HA

7 day programmable room thermostat

The TP75H is a 7 day, programmable room thermostat designed specifically for us in high-current applications such as direct electric heating. The TP75H is capable of switching loads of up to 16A directly without the use of an external load contactor.

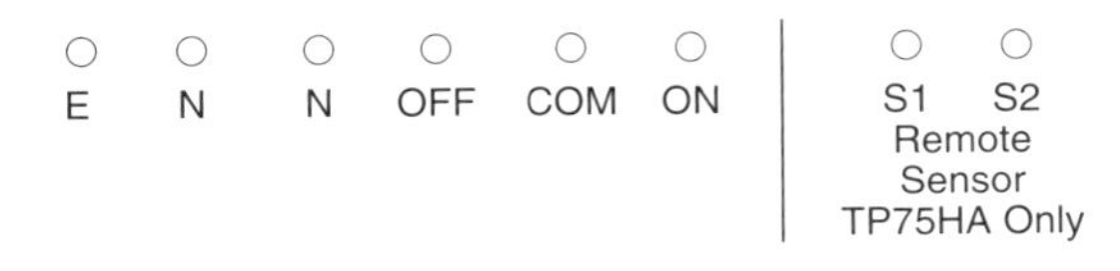

Scale 5–30°C
H88 × W135 × D43
Switch rating 16A (4A)
Up to 6 temperature changes per day
Batteries required 2 × AA

Danfoss-Randall TP5000

Electronic programmable room thermostat

A	B	C
○	○	○
Satisfied	Com	Demand

Scale 5–30°C
H88 × W110 × D28
Switch rating 6A (2A)
Up to 6 temperature changes per day

Danfoss-Randall TP5000-RF

Electronic programmable room thermostat

A	B	C	1	2	3	4	5	6
○	○	○	○	○	○	○	○	○
	N	L						
	MAINS			COM	ZONE 1 ON	ZONE 1 OFF	ZONE 2 ON	ZONE 3 ON

Scale 5–30°C
H88 × W110 × D28
Switch rating 6A (2A)
Up to 6 temperature changes per day

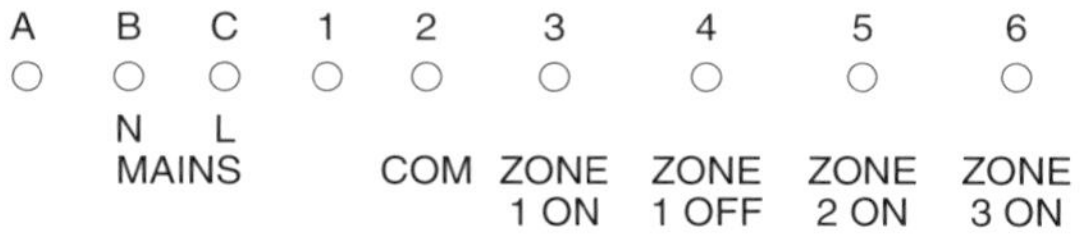

Danfoss-Randall WP75H

5/2 day or 7 day programmable hot water thermostat

						S1	S2
○	○	○	○	○	○	○	○
E	N	N	OFF	COM	ON	Remote Sensor	

Scale 35–65°C
H88 × W135 × D43
Switch rating 16A (4A)
On/offs × 2
Batteries required 2 × AA

Drayton Digistat RF 2

Electronic wireless 24 hour programmable thermostat

N	L	1	2	3
○	○	○	○	○
MAINS		COM	SAT	DEM

Scale 5–30°C
H87 × W87 × D33
Switch rating 2A (1A)
On/offs × 2
Batteries required 4 × AA

Drayton Digistat RF 3

Electronic wireless 24 hour programmable thermostat

7 day version of Digistat RF 2

Eberle RTR-UTQ 9200 RTR-UWQ 9400E

Electromechanical clock thermostat

This is a room thermostat with a built-in time switch which reduces temperature at selected times. It incorporates a three-position switch – clock, moon and sun. With the switch in the clock position, the time switch will automatically switch from set temperature to reduced temperature by 2–10°C. In the moon position the time switch will override the clock to give continuous set-back. In the sun position, the time switch will override the clock to give continuous temperature as set on the thermostat.

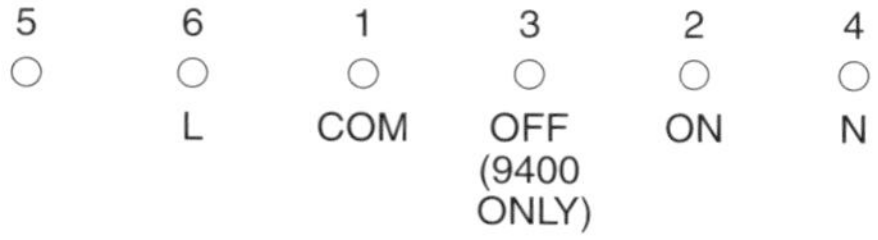

5	6	1	3	2	4
○	○	○	○	○	○
	L	COM	OFF (9400 ONLY)	ON	N

Link 1–6 if required

Scale 5–30°C
H71 × W142 × D32
Switch rating 10A (4A)
On/offs × 3

Eberle RTR-UTQ 9230

24V electromechanical clock thermostat

This is a room thermostat with a built-in time switch which reduces temperature at selected times. It incorporates a three-position switch – clock, moon and sun. With the switch in the clock position, the time switch will automatically switch from set temperature to reduced temperature by 2–10°C. In the moon position, the time switch will override the clock to give continuous set-back. In the sun position, the time switch will override the clock to give continuous temperature as set on the thermostat.

Link 1–6 if required

Scale 5–30°C
H71 × W142 × D32
Switch rating 1A (1A)
On/offs × 3

Eberle RTR-UQW 9300

Electromechanical clock thermostat

7 day version of RTR-UTQ 9200 with 9 on/offs.

Grasslin RTC 7

Programmable room thermostat

1 Com
2 Satisfied
3 Demand

Scale 8–32°C
H140 × W143 × D33
Switch rating 8A(1A)
Up to 25 temperature changes per day
Batteries required 2 × AA

Honeywell CM 31
CM 37

24 hour programmable room thermostat
7 day programmable room thermostat

A LIVE IN
B DEMAND
C SATISFIED

Analogue thermostats with traditional clock face

Scale 5–30°C
H90 × W135 × D29
Switch rating 8A
On/off × 48 per day – CM31
On/off × 6 per day – CM37
Batteries required 2 × AA

Honeywell CM 41

Electronic programmable thermostat

A LIVE IN
B LOAD

Scale 5–30°C
H80 × W130 × D37
Switch rating 8 (3A) SPST
4 time/temperature pairs per day
Batteries required 2 × AA

Honeywell CM 51

7 day electronic programmable thermostat

A LIVE IN
B LOAD

Scale 5–30°C
H80 × W130 × D37
Switch rating 8 (3A) SPST
6 time/temperature pairs per day
Batteries required 2 × AA

Honeywell CM 61
CM 67

24 hour programmable room thermostat
7 day programmable room thermostat

A	B	C
○	○	○
LIVE IN	DEMAND	SATISFIED

Scale 5–30°C
H86 × W130 × D30
Switch rating 8A
On/off × 3 to 12 cycles per hour
Batteries required 2 × AA

Honeywell CM 61 RF
CM 67 RF

Wire free version of CM61, CM67

Honeywell CM 4000
CM 5000

Electronic programmable thermostat

The CM 4000/5000 allows switching times and room temperature to be selected from a central point. An inbuilt sensor is used to detect and control room temperature. The CM 4000 has a 24 hour programming and the CM 5000 has 7 day programming.

A	B	C
○	○	○
L(COM)	ON	OFF

As the unit is battery operated no neutral is required

Scale 8–32°C
H82 × W131 × D35
Switch rating 3A (3A)
Setting periods × 6
Batteries required 2 × AA

Horstmann
Centaurstat 1
Centaurstat 7

Electronic programmable thermostat

The Centaurstat allows switching times and room temperatures to be selected from a central point. An inbuilt sensor is used to select and control room temperature. The Centaurstat 1 has 24 hour setting and the Centaurstat 7 has weekday/weekend setting.

4	○	SPARE
3	○	OFF
2	○	ON
1	○	COMMON

As the unit is battery operated no neutral is required

Scale 6–30°C
H71 × W142 × D30
Switch rating 8A (3A)
Setting periods × 4
Batteries required 3 × AA

Landis & Gyr RAV 1 Chronogyr

Electromechanical clock thermostat

The RAV 1 is a room temperature controller with adjustable automatic night set-back. Two knobs are provided for setting the day and night room temperatures and a sychronous time switch for adjustment of the changeover from day/night temperature times.

1	2	3	4	5	6
○	○	○	○	○	○
L	N	COM	ON	N	E
MAINS					

Link 1–3 if required. Link 2–5

Scale 5–35°C
H75 × W153 × D70
Switch rating 4A (2.5A)

Landis & Gyr RAV 10U Chronogyr RAV 10Ure 24V

Electromechanical clock thermostat

The RAV 10 is a room thermostat with a built-in time switch which reduces temperature at selected times. It incorporates a three-position switch – clock, moon and sun. With the switch in the 'clock' position, the time switch will automatically switch from set temperature to reduced temperature by 2–12°C. In the 'moon' position, the time switch will override the clock to give continuous set-back. In the 'sun' position, the time switch will override the clock to give continuous temperature as set on the thermostat.

1 6 7 5 2 3 4
L L N N OFF ON
LINK LINK

Scale 5–30°C
H80 × W160 × D28
Switch rating 6A

Landis & Gyr RAV 91 Chronogyr

Electromechanical clock thermostat

The RAV 91 is a room thermostat with a built-in time switch which reduces temperature at selected times. It incorporates a three-position switch – clock, moon and sun. With the switch in the 'clock' position, the time switch will automatically switch from set temperature to reduced temperature by 2–12°C. In the 'moon' position, the time switch will override the clock to give continuous set-back. In the 'sun' position, the time switch will override the clock to give continuous temperature as set on the thermostat. The unit has a quartz clock and is powered by batteries.

Q2 Q1
ON COM

Scale 8–27°C
Switch rating 6A (2.5A)

Landis & Gyr REV 10 Chronogyr

Electronic programmable room thermostat

The REV 10 is a room thermostat with a built-in time switch which reduces temperature at selected times. It incorporates a five-position switch – A, B, sun, moon and off. With the switch at A or B the function will be auto operation to heating programme A (two set-back cycles per 24 hours) or B (one set-back cycle per 24 hours). In the sun position, the time switch will override the clock to give continuous temperature as set on the thermostat, and the moon position will give continuous temperature at the pre-set reduced temperature. In the off position the heating will be off unless the frost protection facility, set at 5°C, takes over. The REV 10 has 24 hour setting facility.

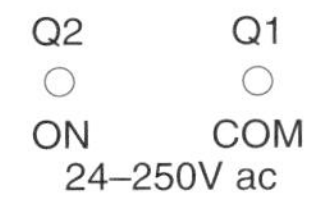

Scale 3–29°C
H89 × W115 × D25
Switch rating 10A (5A)
Setting periods up to 2 set-backs per day
Batteries required 2 × AA

Landis & Gyr REV 20 Chronogyr — Electronic programmable room thermostat

The REV 20 is a room thermostat with a built-in time switch which reduces temperature at selected times. It incorporates a five-position switch – A, B, sun, moon and off. With the switch at A or B the function will be auto operation to heating programme A (two set-back cycles per 24 hours) or B (one set-back cycle per 24 hours). In the sun position, the time switch will override the clock to give continuous temperature as set on the thermostat, and the moon position will give continuous temperature at the pre-set reduced temperature. In the off position the heating will be off unless the frost protection facility, set at 5°C, takes over. The REV 20 has 7 day setting facility.

Q1	Q2	Q3
○	○	○
COM	ON	OFF

24–250V ac

Scale 3–29°C
H90 × W115 × D32
Switch rating 5A (2A)
On/off × 3 per day
Batteries required 3 × AA

Landis and Staefa REV 11 Chronogyr — Electronic programmable room thermostat

As REV 15 with daily operation

Landis & Staefa REV 15/15T Chronogyr — Electronic programmable room thermostat

The REV 15 incorporates a self-learning control algorithm that enables the thermostat to adjust to local climate, building and heating installation environments to provide optimum comfort benefits. The REV 15T has a remote operation facility. Both units have 5/2 day operating modes and are battery powered.

L	L1
○	○
COM	DEM

Scale 0–40°C
H104 × W128 × D37
Switch rating 8A (3.5A)
Batteries required 2 × AA

Landis & Staefa REV 22/22T Chronogyr — Electronic programmable room thermostat

The REV 22 incorporates a self-learning control algorithm that enables the thermostat to adjust to local climate, building and heating installation environments to provide optimum comfort benefits. The REV 22T has a remote operation facility. Both units have 7 day programming and holiday programming with three comfort settings per day. Also incorporates night set-back and frost protection. The units are battery powered.

L	L1	L2
○	○	○
COM	N.O.	N.C.

Scale 0–40°C
H104 × W128 × D37
Switch rating 6A (2.5A)
Batteries required 2 × AA

Potterton PET 1

Electronic programmable thermostat

The PET 1 allows switching times and room temperatures to be selected from a central point. An inbuilt sensor is used to detect and control room temperature. The PET 1 has weekday/weekend programming facility and low-limit frost protection override.

2	L	1	N	E
○	○	○	○	○
OFF	L(COM)	ON	N	E

Scale 6–29°C
H88 × W142 × D46
Switch rating 6A (1A)
Setting periods × 4 in 24 hours
Batteries required 4 × AA

Randall TP 1

As Potterton Pet 1

Randall TP 2, TP 3, TP 4, TP 5

Electronic programmable thermostat

The TP 2-5 allows switching times and room temperatures to be selected from a central point. An inbuilt sensor is used to detect and control room temperatures. The TP 2 and TP 4 have 24 hour programming and the TP 3 and TP 5 have weekday/weekend programming.

3	2	1
○	○	○
ON	OFF	COM

As the unit is battery powered no neutral is required

Scale TP 2 and TP 3 16–30°C
Scale TP 4 and TP 5 5–30°C
H81 × W98 × D34
Switch rating 6A
Setting periods × 6 in 24 hours
Batteries required 2 × AA

Randall TP 6, TP 7

Electronic programmable thermostat with timed hot water Basic/Full control

The PT 6 and TP 7 allow switching times and room temperatures to be controlled from a central point. An inbuilt sensor is used to detect and control room temperatures. The hot water temperature can be controlled using a conventional cylinder thermostat. For Basic control, remove white socket from back of unit.

E	N	L	1	2	3	4	5	6
○	○	○	○	○	○	○	○	○
E	N	L	HW	HW	HW	CH	CH	CH
MAINS			ON	COM	OFF	ON	COM	OFF

Scale TP 6 16–30°C
Scale TP 7 5–30°C
H105 × W150 × D38
Switch rating 3A
CH setting periods × 6 in 24 hours
HW on/offs × 2

Randall TP 8, TP 9

Electronic programmable thermostat with timed hot water Basic/Full control

The TP 8 and TP 9 allow switching times and room temperatures to be controlled from a central point. A remote sensor is used to detect and control room temperature. The hot water can be controlled using a conventional cylinder thermostat. For Basic control, remove white socket from back of unit.

E	N	L	1	2	3	4	5	6	7	8
○	○	○	○	○	○	○	○	○	○	○
E	N	L	HW	HW	HW	CH	CH	CH	SENSOR	
	MAINS		ON	COM	OFF	ON	COM	OFF		

Link L–2–5 if required

Scale TP 8 16–30°C
Scale TP 9 5–30°C
H105 × W150 × D38
Switch rating 3A
CH setting periods × 6 in 24 hours
HW on/offs × 2

Smiths Centroller 2000

Electronic 24 hour Basic/Full programmer

This programmer was supplied with room and cylinder sensors. The on/off times are divided up into day/nighttime periods and room and water temperature are set on the programmer itself.

Two diagrams are shown; note that boiler and pump switching were done by the programmer and not the auxiliary switches of motorized valves.

H125 × W170 × D70
Switch rating 5A (2A)

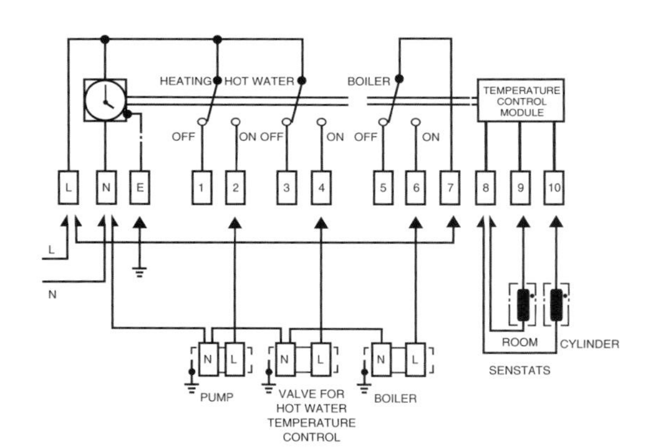

Figure 3.1 *Centroller 2000/3000. Basic system pumped heating gravity hot water. Do NOT use on low water content boilers.*

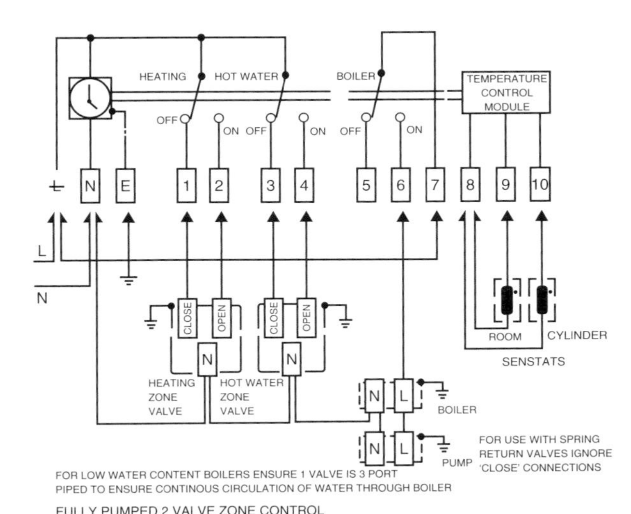

Figure 3.2 *Centroller 2000/3000. Fully pumped 2 valve zone control. For spring return motorized valves ignore terminals 1 and 3.*

Smith Centroller 3000

Electronic 24 hour Basic/Full programmer

As Centroller 2000 with boost facility for hot water.

Smiths ERS 1

Electronic programmable thermostat

The ERS 1 allows switching times and room temperatures to be selected from a central point. An inbuilt sensor is used to detect and control room temperature. The ERS 1 has weekday/weekend programming facility and low-limit frost protection override.

2	L	1	N	E
○	○	○	○	○
OFF	L(COM)	ON	N	E

Scale 6–29°C
H88 × W142 × D46
Switch rating 6A (1A)
Setting periods × 4 in 24 hours
Batteries required 4 × AA

Sunvic EC 1401/1451

Electronic clock thermostat

The Sunvic EC electronic clock thermostat is designed to control room temperature at two pre-selected levels – day and night. During the On periods the temperature is controlled by the selected day temperature and at other times controlled to the night temperature setting.

Voltage-free switching unless L–5–6 linked

Scale 10–40°C
H110 × W180 × D65
Switch rating 5A (1A)

Sunvic TLC 2358

Electromechanical clock thermostat

The TLC is a room thermostat with a built-in time switch which reduces temperature at selected times. It incorporates a three-position switch – clock, moon and sun. With the switch in the clock position, the time switch will automatically switch from set temperature to reduced temperature by 6°C. In the moon position, the time switch will override the clock to give continuous set-back. In the sun position, the time switch will override the clock to give continuous temperature as set on the thermostat.

1 ON, 2 OFF, 3 L(COM), 4 N
MAINS

Scale 3–27°C
H87 × W157 × D47
Switch rating 2A (1A)

Sunvic TLX 6501

Programmable room thermostat

1 COM, 2 DEMAND, 3 SATISFIED

Scale 5–40°C
H82 × W120 × D31
Switch rating 6A (2A)
Up to 6 temperature changes per day
Batteries required 2 × AA

Switchmaster Serenade

Electronic programmable room thermostat

The Serenade allows switching times and room temperature to be selected from a central point. An inbuilt sensor is used to detect and control room temperature. The Serenade has 7 day programming and a frost protection facility.

4 ON, 5 COM, 6 OFF

As the unit is battery operated no neutral is required

Scale 5–29.5°C
H95 × W161 × D40
Switch rating 3A (2A)
Setting periods × 3
Batteries required 3 × AA

Switchmaster Symphony

Electronic 7 day Full programmer

The Symphony is supplied with room and cylinder sensors.

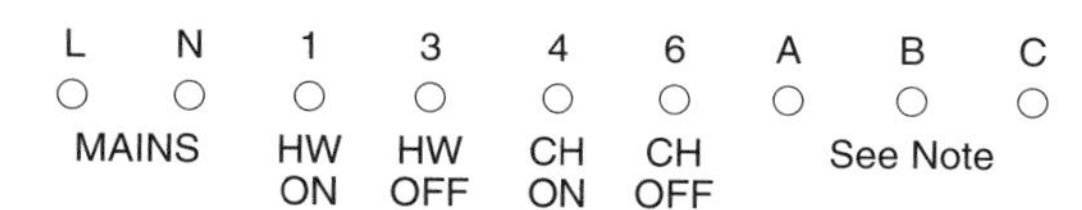

A–C Room sensor. B–C Cylinder sensor

On/off × 3
H95 × W161 × D40
Switch rating 3A

Vaillant CRT 394 Calotrol

Electromechanical clock thermostat

The VRT 394 is a room thermostat with a built-in time switch which reduces temperature at selected times. It incorporates a three-position switch – clock, moon and sun. With the switch in the clock position, the time switch will automatically switch from set temperature to reduced temperature by 5°C. In the moon position, the timeswitch will override the clock to give continuous set-back. In the sun position, the timeswitch will override the clock to give continuous temperature as set on the thermostat. When tappets are pushed in, the clock functions at full set temperature, and when tappets remain out, the clock functions at reduced set temperature.

Scale 5–30°C
H75 × W146 × D28
Switch rating 10A

Voltage free switching unless 2–3 linked

The VRT 394 can also function in an on/off mode when it will turn off a boiler, e.g. during the nighttime period. For this facility wire as below:

Voltage-free switching unless 2–7 linked

Vaillant VRT QT4 Calotrol
VRT QW4 Calotrol

Electromechanical clock thermostat

The QT has a 24 hour clock and the QW has a 7 day clock. They incorporate a three-position switch – clock, moon and sun. With the switch in the clock position, the time switch will automatically switch from set temperature to reduced temperature by 6°C. In the moon position, the time switch will override the clock to give continuous set-back. In the sun position, the time switch will override the clock to give continuous temperature as set on the thermostat. The set-back of 5°C can be altered up to 10°C by first isolating the power supply and removing the casing by releasing bottom screws and tilting case upwards. Adjust the potentiometer to required level and re-assemble case. When tappets are pushed in, the clock functions at full set temperature, and when tappets remain out, the clock functions at reduced set temperature.

4	1	2	3
○	○	○	○
ON	N	L	COM
	MAINS		

Scale 5–30°C
H75 × W142 × D35
Switch rating 2A

Voltage-free switching unless 2–3 linked

Vaillant VRT-UT2-394 240V Calotrol VRT-UT2-396 24V Calotrol

Electromechanical clock thermostat

The VRT-UT2-394/396 are room thermostats with a built-in time switch which reduces temperature at selected times. They incorporate a three-position switch – clock, moon and sun. With the switch in the clock position, the time switch will automatically switch from set temperature to reduced temperature by 5°C. In the moon position, the timeswitch will override the clock to give continuous set-back. In the sun position, the timeswitch will override the clock to give continuous temperature as set on the thermostat. When tappets are pushed in the clock functions at full set temperature and when tappets remain out the clock functions at reduced set temperature.

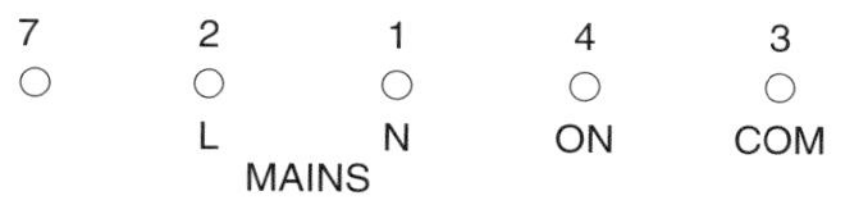

7	2	1	4	3
○	○	○	○	○
	L	N	ON	COM
	MAINS			

Scale 5–30°C
H75 × W146 × D28
Switch rating 10A

Voltage-free switching unless 2–3 linked

The VRT-UT2-394/396 can also function in an on/off mode when it will turn off a boiler, e.g. during the nighttime period. For this facility wire as below:

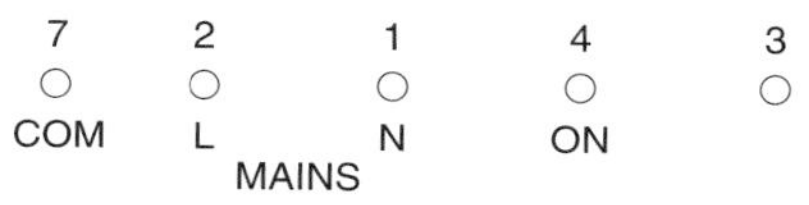

7	2	1	4	3
○	○	○	○	○
COM	L	N	ON	
	MAINS			

Voltage-free switching unless 2–7 linked

4

Cylinder and pipe thermostats

The cylinder thermostat is a device for detecting and setting the temperature of water in the domestic hot water cylinder. It should be in a position so that the householder can easily make any adjustments that may be required. If possible, it is best to site it away from the flow/return pipes to avoiding detecting conducted heat. The thermostat is usually located a third of the way up the cylinder and virtually all are clamped to the cylinder with a metal band or spring wire. The exceptions to this rule are the Potterton PTT1 and PTT2. On these, the actual thermostat is fixed on a convenient wall and a pre-wired probe is attached to the cylinder. The probes are available with 2m or 10m leads. These thermostats are ideal where a cylinder is located in a loft, eaves cupboard or similar difficult-to-get-at location.

Virtually all cylinder and pipe thermostats have SPDT switching and are suitable for all voltages up to 240 V. Some are pre-wired and when using one of these for a SPST application the wire not used must be safely terminated as it will be live when the thermostat is in a satisfied state.

There are two instances where fixing the cylinder thermostat can be problem. These are the horizontally mounted cylinder and the square or oblong copper tank. To deal with the horizontal cylinder first. It is probably going to be a trial and error exercise as to location but a third of the way up as normal would be the best place to start. As for the square or oblong copper tank, one method is to solder two copper tags onto the tank about 12–18 inches apart and a third of the way up. Two pieces of flattened 15mm tube will do the job. Then the thermostat can be fixed using the spring wire method. Alternatively, the Potterton cylinder thermostat with sensor could be used with the sensor being held in position with suitable tape.

Pipe thermostats are usually cylinder thermostats with a modified base. They function just the same and are used for various reasons, including pump overrun thermostats.

ACL HTS

Common	3	Scale 50–80°C
Demand	1	H114 × W58 × D67
Satisfied	2	15A res.

ACL HTS 2

Common	Red	Scale 50–80°C
Demand	Black	H110 × W34 × D40
Satisfied	Yellow	3A res.

1.5m pre-wired.

ACL HTS 2/S

As HTS 2 but metal strap fixing.

ACL HTS 3

Common	C	Scale 50–80°C
Demand	1	H100 × W40 × D45
Satisfied	2	3A res.

Barlo CT 1

Common	Red	Scale 50–80°C
Demand	Black	H100 × W34 × D40
Satisfied	Yellow	3A res.

Benefit

Common	1	Scale 15–90°C
Demand	2	H116 × W50 × D54
Satisfied	3	15A res. (2.5A)

Danfoss ATC

Common	C	Scale 20–90°C
Demand	NC	H100 × W60 × D57
Satisfied	NO	15A res. (2.5A ind.)

Danfoss ATF

Pipe frost thermostat

Common	C	Scale 5–90°C
Demand	NC	H100 × W60 × D57
Satisfied	NO	6A

Danfoss ATP

Pipe thermostat

Common	C	Scale 30–90°C
Demand	NC	H100 × W60 × D57
Satisfied	NO	6A

Drayton CS 1

Common	1	Scale 20–90°C
Demand	2	H92 × W60 × D59
Satisfied	3	6A res. (2A ind.)

Drayton CS 2

Common	1	Scale 20–90°C
Demand	2	H90 × W40 × D45
Satisfied	4	15A res. (2.5A)

Eberle rar

Common	1	Scale 15–90°C
Demand	2	H116 × W50 × D54
Satisfied	3	15A res. (2.5A)

Honeywell L641A

Common	C	Scale 50–80°C
Demand	1	H79 × W44 × D44/54
Satisfied	2	4A res. (2A ind.)

Honeywell L641B

Pipe frost thermostat

Common	C	Scale 10–40°C
Demand	1	H79 × W44 × D56
Satisfied	2	4A (2A)

Honeywell L697A

Common	1	Scale 100–180°F
Demand	2	H120 × W94 × D50
Satisfied	3	20A res.

Honeywell L6090A

Common	C	Scale 30–90°C
Demand	1	H95 × W50 × D85
Satisfied	2	6A res. (4A ind.)

Honeywell L6190B

Common	C	Scale 25–95°C
Demand	1	H92 × W48 × D77
Satisfied	2	10A res. (2.5A)

Horstmann HCT – 1

Common	1	Scale 15–90°C
Demand	2	H116 × W50 × D54
Satisfied	3	15A res. (2.5A)

Landis & Gyr RAM 1

Common	1	Scale 15–90°C
Demand	2	H116 × W50 × D54
Satisfied	3	15A res. (2.5A)

Landis & Gyr RAM 21

Common	1	Scale 50–80°C
Demand	2	H141 × W50 × D42
Satisfied	3	6A res. (3.5A)

Potterton PTT 1

Common	L	Scale 45–75°C
Demand	H	H78 × W70 × D40
Satisfied	C	5A res. (2.5A)
Neutral	N	

This thermostat is fitted remote from the cylinder using a probe supplied with a 2m lead. A 10m lead is available. The thermostat is also supplied with indicator neons to show whether temperature is reached. The wiring of a neutral is essential.

Potterton PTT 2

From time control	TL	
Perm live/Common	L	Scale 45–75°C
Demand	H	H78 × W70 × D40
Satisfied	C	5A res. (2.5A)
Neutral	N	

This thermostat has a boost facility and a permanent live is required. It is fitted remote from the cylinder using a probe supplied with a 2m lead. A 10m lead is available. The thermostat is also supplied with indicator neons to show whether temperature is reached. The wiring of a neutral is essential.

Potterton PTT 100

Common	TL	Scale 30–90°C
Demand	H	H110 × W52 × D70
Satisfied	C	16A (2A)

Proscon SOA

Common	3	Scale 33–83°C
Demand	1	H92 × W57 × D32
Satisfied	2	3A res. (2A ind.)

Randall CN4

Common	1	Scale 15–90°C
Demand	2	H116 × W50 × D54
Satisfied	3	15A res. (2.5A)

Smiths C

Common	1	Scale 120–180°F
Demand	2	H118 × W53 × D40
Satisfied	3	3A res.

Sopac SA 0570

Common	C	Scale 5–70°C
Demand	1	H100 × W50 × D40
Satisfied	2	16A res. (2.5A)

Sopac SA 0701

Common	C	Scale 25–90°C
Demand	1	H101 × W38 × D38
Satisfied	2	16A res. (2.5A)

Sopac SA 2590

Common	C	Scale 25–90°C
Demand	1	H100 × W50 × D40
Satisfied	2	16A res. (2.5A)

Sunvic PA 2252

Pipe thermostat

Common	3	Scale 40–90°C
Demand	1	H127 × W53 × D62
Satisfied	2	15A res.

Sunvic SA 36

Common	3	Scale 130–180°F
Demand	1	H105 × W55 × D63
Satisfied	2	15A res.

Sunvic SA 1452

Common	3	Scale 46–90°C
Demand	1	H100 × W50 × D50
Satisfied	2	6A res. (2.5A)

Sunvic SA 1453

As SA 1452 but pre-wired.

Common	Brown	Scale 46–90°C
Demand	Black	H100 × W50 × D50
Satisfied	Blue	6A res. (2.5A)

Sunvic SA 1502

As SA 1452, pre-wired with 4-way plug for duoplug or clock box systems. The colour coding of the flex is:

Common	Blue
Demand	Yellow
Satisfied	White

Sunvic SA 1503

As SA 1502.

Sunvic SA 2401

Common	3	Scale 40–90°C
Demand	1	H109 × W53 × D59
Satisfied	2	15A res.

Sunvic SA 2451

Common	3	Scale 40–90°C
Demand	1	H109 × W53 × D59
Satisfied	2	15A res.

Sunvic SA 2452

Common	3	Scale 40–90°C
Demand	1	H95 × W33 × D42
Satisfied	2	5A (1A)

Sunvic SA 2453

As SA 2452 but pre-wired.

Common	Blue	Scale 40–90°C
Demand	Yellow	H95 × W33 × D42
Satisfied	White	5A (1A)

Sunvic SA 2501

As 2451, but pre-wired with 4-way plug fitted for duoplug system. Colour coding as SA 1502.

Sunvic SA 2502

As SA 2453 with plug for duoflow system.

Sunvic SA 2601

Common	3	Scale 45–85°C
Demand	1	H95 × W33 × D42
Satisfied	2	5A (1A)

Switchmaster SCT 1

Common	1	Scale 40–80°C
Demand	2	H114 × W45 × D50
Satisfied	3	6A res. (1A)

Teddington FEA

Common	1	Scale 40–80°C
Demand	2	H114 × W45 × D50
Satisfied	3	6A res. (1A)

Tower CS

Common	Red	Scale 50–80°C
Demand	Black	H100 × W34 × D40
Satisfied	Yellow	3A res.

1.5m pre-wired.

Trac TS 30

Common	1	Scale 15–90°C
Demand	2	H116 × W50 × D54
Satisfied	3	15A res. (2.5A)

Wickes CS

Common	1	Scale 50–80°C
Demand	2	H141 × W50 × D42
Satisfied	3	6A res. (3.5A)

5

Room, frost and low-limit thermostats

Room thermostats come in many shapes and sizes with different switching arrangements, current ratings and facilities such as neon indicators, off position, locking device, thermometer, night setback, etc. However, its function, when used with domestic central heating, is to act as a temperature-operated switch to turn off the pump, close a motorized valve, etc.

As the thermostat is designed to operate on fluctuations in air temperature, it must be sited in a position where there is good air movement and circulation. This would normally be the hall or largest living room, about 5 feet (1.5m) from the floor. It is also necessary to ensure that it is influenced by a radiator not fitted with a thermostatic radiator valve, but it must not be sited immediately adjacent to it.

Some examples of where **not to** site a room thermostat are:

(a) within 6 inches of an internal corner as air circulates – a corner of a room is regarded as dead air space

(b) over or near an artificial heat source, e.g. table lamp, television

(c) in a kitchen or cupboard

(d) in a room containing an open fire, gas fire, electric fire, or similar heating appliance not influenced by the thermostat

(e) on an outside wall

(f) on an airing cupboard wall

(g) on a chimney breast that may be used

(h) in direct sunlight

(i) in a draught

(j) behind curtains.

Having selected the correct position, which is often a matter of compromise, ensure that the thermostat is wired up correctly. This is straight forward but it is essential for greater accuracy to wire the live into the correct terminal so that the heat anticipator, if fitted, functions correctly.

One other thing to remember is that on 24V room thermostats there is usually an adjustment to be made inside the thermostat so that the anticipator scale coincides with the current rating of the gas valve of the boiler or warm air heater. A 24V thermostat must not be used on a 240V supply, but a 240V room thermostat can be used on a 24V system, although it may perform poorly and such practice is not to be encouraged.

If part of the system is likely to suffer damage from freezing in cold weather, it is necessary to fit a frost or low-limit thermostat. Likewise, protection of the property itself is sometimes the requirement, e.g. timber framed houses may require a low-limit thermostat set at about 11°C to avoid condensation forming. Usually, the frost or low-limit thermostats are wired to override any time clock and room thermostat. Many electricians think that room, frost and low-limit are wired differently because one acts at high temperature and the others at low temperature. However, all three thermostats break on temperature rise and so are wired the same. When wiring a frost thermostat on a gravity hot-water/pumped central heating system, it is necessary to override both hot water and heating channels of a programmer, and so a double-pole thermostat, such as a Sopac TA 147, has to be used. An alternative is the conventional SPST thermostat, with a double-pole relay. Both methods are shown on page 257. Some manufacturers include a frost position on their normal room thermostats or have boxes for fitting the frost thermostat outside on a wall, such as the Sunvic BX3.

ACL DT

Common	3	Scale 5–25°C
Demand	1	H66 × W100 × D39
Satisfied	2	6A res. (2A)
Neutral	N	

ACL ST

Common	3	Scale 5–25°C
Demand	1	H66 × W100 × D39
Satisfied	None	6A res. (2A)
Neutral	N	

ACL TA350

Common	1	Scale 6–30°C
Demand	3	H72 × W72 × D44
Satisfied	2	16A res. (2.5A)
Neutral	None	

ACL TS142

Common	1	Scale 3–30°C
Demand	2	H71 × W71 × D35
Satisfied	None	16A res. (4A)
Neutral	4	

ACL-Drayton RTS 1 and 2

Electronic thermostat, 240V only

Common	L	Scale 10–30°C
Demand	3	H86 × W86 × D37
Satisfied	None	2A (1A)
Neutral	N	RTS2 has LED 'ON' indicator

ACL-Drayton RTS 3

Frost thermostat, 240V only

Common	L	Scale 3–10°C
Demand	3	H86 × W86 × D37
Satisfied	None	2A (1A)
Neutral	N	

ACL-Drayton RTS 4

Electronic thermostat

Perm live	L	Scale 10–30°C
Common	1	H86 × W86 × D37
Demand	3	2A (1A)
Satisfied	2	
Neutral	N	

It is essential that a 240V supply is wired to L and N. Link L–1 if voltage-free contacts are not required.

ACL-Drayton RTS 5

Optimum start and save facility

Common	1	Scale 10–30°C
Demand	3	H86 × W86 × D37
Satisfied	None	2A (1A)
Neutral	N	

Barlo RT 1

Common	1	Scale 3–30°C
Demand	2	H71 × W71 × D35
Satisfied	None	16A res. (4A)
Neutral	4	

Benefit BRFT 10

Common	1	Scale 5–30°C
Demand	2	H68 × W90 × D40
Satisfied	3	16A res. (2.5A)
Neutral	None	

Brassware Ferroli

Common	2	Scale 5–30°C
Demand	1	H78 × W78 × D36
Satisfied	3	16A res. (2.5A)
Neutral	None	

Danfoss HCT

Common	3	Scale 5–30°C
Demand	5	H88 × W135 × D32
Satisfied	6	2A (1A)
Remote sensor	A–B	Batteries 2 × AA

Danfoss-Randall HT

Incorporates day/night switch

Common	3	Scale 5–30°C
Demand	5	H88 × W135 × D32
Satisfied	6	2A (1A)
Neutral		

Danfoss-Randall RET 230 C, L, NA

Electronic thermostat

Common	L	Scale 5–30°C
Demand	3	H86 × W85 × D42
Satisfied	4	3A (1A)
Neutral	N	

Danfoss-Randall RET-B Battery powered

Common	3	Scale 5–30°C
Demand	2	H88 × W110 × D28
Satisfied	1	6A (2A)
Neutral	None	

Danfoss RMT 24 – 24 V

Common	1	Scale 8–30°C
Demand	2	H80 × W80 × D40
Satisfied	3	10A (4)
Neutral	4	Neutral is – of circuit

Danfoss RMT 24T

With night set-back facility/thermometer

Common	1	Scale 8–30°C
Demand	2	H80 × W80 × D40
Satisfied	3	10A (4A)
Neutral	4 + 5	Neutral is – of circuit
NSB	6	

Dansfoss RMT 230

Common	1	Scale 8–30°C
Demand	2	H80 × W80 × D40
Satisfied	3	10A (4A)
Neutral	4	

Danfoss RMT 230T

With night set-back facility/thermometer

Common	1	Scale 8–30°C
Demand	2	H80 × W80 × D40
Satisfied	3	10A (4A)
Neutral	4	
NSB	switch 5 and 6	

Danfoss RTF

Frost thermostat

Common	1	Scale − 5°C fixed
Demand	2	H80 × W80 × D40
Satisfied	None	16A (2.5A)
Neutral	None	

Danfoss RT1

Digital electronic thermostat

Common	1	Scale 5–30°C
Demand	3	H81 × W98 × D34
Satisfied	2	6A SPDT
Neutral	None	Battery powered

Danfoss-Randall RT 1 – RF

As RT 1 with remote sensor.

Danfoss-Randall RT 2

As RT 1 with integral timer.

Danfoss-Randall RT 2 – RF

As RT 2 with remote sensor.

Danfoss-Randall RT 51

Manual day/night feature

Common	B	Scale 5–30°C
Demand	C	H88 × W110 × D28
Satisfied	A	6A (2A)
Neutral		Batteries 2 × AA

Danfoss-Randall RT 51 – RF

As RT 51 with remote sensor, RX1, RX2 or RX3, see TP5E-RF.

Danfoss ST

Common	2	Scale 40–80°F
Demand	3	H112 × W53 × D53
Satisfied	1	
Neutral	None	0–2 res.

Danfoss TWE

Common	2	Scale 0–30°C
Demand	1	H82 × W82 × D33
Satisfied	None	10A res. (6A)
Neutral	4	

Danfoss TWK

Common	2	Scale 0–30°C
Demand	1	H82 × W82 × D33
Satisfied	3	10A res. (6A)
Neutral	4	

Danfoss TWL – 24 V

Common	1	Scale 5–30°C
Demand	4	H82 × W82 × D33
Satisfied	None	1A res. (1A)
Neutral	None	

Danfoss TWLT – 24 V

Night set-back facility

Common	C	Scale 5–30°C
Demand	4	H82 × W82 × D33
Satisfied	None	1A res. (1A)
Neutral	None	
NSB	6	

Danfoss TWP

Common	1	Scale 5–30°C
Demand	2	H82 × W82 × D33
Satisfied	3	10A res. (4A)
Neutral	4	

Danfoss TWPT

Night set-back facility

Common	1	Scale 5–30°C
Demand	2	H82 × W82 × D33
Satisfied	3	10A res. (4A)
Neutral	4	
NSB	5	

Danfoss TWR

Common	2	Scale 0–30°C
Demand	1	H82 × W82 × D33
Satisfied	None	10A res. (6A)
Neutral	4	

Danfoss TWR 24 – 24 V

Common	2	Scale 5–30°C
Demand	1	H82 × W82 × D33
Satisfied	None	1A res.
Neutral	None	

Drayton Digistat 1

Electronic thermostat

Common	1	Scale 5–30°C
Demand	3	H88 × W88 × D33
Satisfied	2	2A (1A)
Neutral	None	

Batteries required 2 × AA

Drayton Digistat RF 1

Electronic wireless digital thermostat

Used in conjunction with SCR receiver

Common	1	Scale 5–30°C
Demand	3	H88 × W88 × D33
Satisfied	2	2A (1A)
Neutral	N	Batteries required 2 × AA
Live	L	

Drayton Roomstat

Common	1	Scale 5–30°C
Demand	2	H81 × W81 × D31
Satisfied	3	10A res.
Neutral	4	

Drayton RT°C

Common	3	Scale 2–27°C
Demand	1	H66 × W100 × D39
Satisfied	2	6A res. (2A)
Neutral	N	

Drayton RTE

Common	1	Scale 5–30°C
Demand	2	H79 × W79 × D27
Satisfied	3	10A res. (4A)
Neutral	4	

Drayton RTL – 24V

Common	3	Scale 13–24°C
Demand	1	H66 × W100 × D39
Satisfied	2	6A res. (2A)
Neutral	None	

Drayton RTM

Common	3	Scale 13–24°C
Demand	1	H66 × W100 × D39
Satisfied	2	6A res. (2A)
Neutral	N	

Eberle RTR3521

Common	1	Scale 3–30°C
Demand	2	H71 × W71 × D35
Satisfied	None	16A res. (4A)
Neutral	4	

Eberle RTR6121

Common	1	Scale 5–30°C
Demand	2	H79 × W79 × D27
Satisfied	None	10A res. (4A)
Neutral	4	

Ecko ET

Common	L	Scale 0–30°C
Demand	H	H78 × W133 × D41
Satisfied	C	20A res.
Neutral	N	

Ekco ETS

With on/off switch

Common	L	Scale 0–30°C
Demand	H	H78 × W133 × D41
Satisfied	C	20A res.
Neutral	N	

Ekco ET 16

Common	3	Scale 3–27°C
Demand	1	H90 × W86 × D46
Satisfied	None	16A res.
Neutral	4	

Ekco RS

Common	4	
Demand	3	H132 × W75 × D65
Satisfied	2	20A res.
Neutral	1	

Honeywell CT200

Electronic thermostat

With night set-back facility

Common	A	Scale 5–30°C
Demand	B	H84 × W84 × D34
Satisfied	C	8 (3A)

Batteries required 2 × AA

Honeywell T87F 24V

2 wire control — Scale 5–30°C
Circular 93mm
1.5A

Honeywell T403A 1018

Common	3	Scale 40–80°F
Demand	1	H98 × W83 × D40
Satisfied	None	2A res.
Neutral	2	

Honeywell T403A 1125

Frost thermostat
With tamper-proof cover

Common	3	Scale 30–70°F
Demand	1	H98 × W83 × D40
Satisfied	None	1A res.
Neutral	None	

Honeywell T403A 1141

Common	3	Scale 40–80°F
Demand	1	H98 × W83 × D40
Satisfied	None	2A res.
Neutral	None	

Honeywell T406

Frost thermostat
With tamper-proof cover

Common	1	Scale 0–20°C
Demand	3	H87 × W114 × D42
Satisfied	None	20A res. (4A)
Neutral		

Honeywell T473X

Common	3	Scale 48–72°F
Demand	2	H85 × W132 × D65
Satisfied	None	17.5A res. (6.5A)
Neutral	None	

Honeywell T498

Common	3	
Demand	1	H115 × W73 × D65
Satisfied	None	22A res.
Neutral	2	

Honeywell T603

Common	1
Demand	3
Satisfied	4
Neutral	None

Honeywell T803 – 24V

Common	3	Scale 42–80°F
Demand	1	H95 × W78 × D35
Satisfied	None	0.18–0.8A res.
Neutral	None	

Honeywell T822 – 24V

2 wire control — Scale 13–35°C
H125 × W79 × D31
0.18–0.8A res.

Honeywell T832 1083 – 24V

Incorporates two temperature setting levers and a hand wound clock, which enables user to select day and night running temperatures.

Circular in shape 93mm.

Honeywell T4160A

Frost thermostat
With tamper-proof cover

Common	1	Scale 0–20°C
Demand	3	H79 × W83 × D49
Satisfied	None	2A res. (2A)
Neutral	None	

Honeywell T4160C

With night set-back facility

Common	1	Scale 10–30°C
Demand	3	H79 × W83 × D49
Satisfied	None	2A res. (2A)
Neutral	None	
NSB	5	

Honeywell T4360A

Frost thermostat

Common	1	Scale 3–22°C
Demand	3	H84 × W84 × D42
Satisfied	None	10A (3A)
Neutral	None	

Honeywell T4360B 1015

Common	1	Scale 10–30°C
Demand	3	H84 × W84 × D42
Satisfied	None	16A
Neutral	None	

Honeywell T4360E

With night set-back facility

Common	1	Scale 10–30°C
Demand	3	H84 × W84 × D42
Satisfied	None	10A res. (3A)
Neutral	2 and 5	
NSB	6	

Honeywell T6060 A/B/C

T6060A – no anticipator
T6060B – with anticipator and optional thermometer
T6060C – with anticipator, night set-back and optional thermometer

Common	1	Scale 10–30°C
Demand	3	H87 × W114 × D42
Satisfied	4	20A res. (4A)
Neutral	2	**(not T6060A)**

Honeywell T6160B

Common	1	Scale 10–30°C
Demand	3	H79 × W83 × D49
Satisfied	4	2A res. (2A)
Neutral	2	

Honeywell T6360B 1028

Common	1	Scale 10–30°C
Demand	3	H84 × W84 × D42
Satisfied	4	10A (3A res.) term 3
Neutral	2	6A (2A res.) term 4

Honeywell T6360B 1036

With indictor lamp illuminated when calling for heat

Common	1	Scale 10–30°C
Demand	3	H84 × W84 × D42
Satisfied	4	10A (3A res.) term 3
Neutral	2	6A (2A res.) term 4

Honeywell T6360B 1069

With tamper-proof cover

Common	1	Scale 10–30°C
Demand	3	H84 × W84 × D42
Satisfied	4	10A (3A res.) term 3
Neutral	2	6A (2A res.) term 4

Honeywell T6360B 1085

Common	1	Scale 1–5°C
Demand	3	H84 × W84 × D42
Satisfied	4	10A (3A res.) term 3
Neutral	2	6A (2A res.) term 4

Horstmann HRT 1

Common	1	Scale 3–30°C
Demand	2	H71 × W71 × D35
Satisfied	None	16A res. (4A)
Neutral	4	

KDG Range

As Sopac/Smiths Industries.
For model see inside cover.

Landis & Gyr RAD 1A

Common	1	Scale 5–30°C
Demand	2	H78 × W78 × D43
Satisfied	3	16A res. (2.5A)
Neutral	4	

Landis & Gyr RAD 1F

Common	1	Scale 5°C fixed
Demand	2	H78 × W78 × D43
Satisfied	3	16A res. (2.5A)
Neutral	None	

Landis & Gyr RAD 1N

Common	1	Scale 5–30°C
Demand	2	H78 × W78 × D43
Satisfied	3	16A res. (2.5A)
Neutral	None	

Landis & Gyr RAD 5

Common	6	Scale 5–30°C
Demand	2	H80 × W80 × D27
Satisfied	None	6A res. (2.5A)
Neutral	4	

Landis & Gyr RAD 7

Common	1	Scale 5–30°C
Demand	2	H82 × W82 × D30
Satisfied	None	15A res. (4A)
Neutral	4	

Nettle

The Nettle range of thermostats is similar to the Honeywell T6060 range.

Pegler SR 2

Common	L	Scale 0–30°C
Demand	H	H78 × W133 × D41
Satisfied	C	20A res.
Neutral	N	

Potterton PRT 1

Common	L	Scale 5–30°C
Demands	H	H78 × W70 × D38
Satisfied	None	5A res. (2.5A)
Neutral	N	

This thermostat has indicators to show whether the set temperature has been reached. The wiring of a neutral is essential.

Potterton PRT 2

Common	TL4	Scale 4–30°C
Demand	H3	H84 × W110 × D38
Satisfied	None	5A res. (2.5A)
Neutral	N5	

If the thermostat is to be used for 240V, link TL to COM. If the thermostat is to be used for switching a different voltage, e.g. 24V, the switch contacts are Terminal COM and H and the above link should not be fitted. However, a 240V supply to Terminal TL must be fitted and a neutral is required in all applications. This thermostat is fitted with indicators to show whether the set temperature has been reached.

Potterton PRT 100 DT

Common	TL	Scale 5–30°C
Demand	H	H70 × W70 × D31
Satisfied	C	10A (3A)
Neutral	N	

Potterton PRT 100 FR

Frost thermostat

Common	L	Scale −10–15°C
Demand	H	H70 × W70 × D31
Satisfied	C	10A (3A)
Neutral	N	

Potterton PRT 100 ST

Common	TL	Scale 5–30°C
Demand	H	H70 × W70 × D31
Satisfied	None	10A (3A)
Neutral	N	

Proscon LC

Common	3	Scale 5–25°C
Demand	2	H113 × W67 × D51
Satisfied	1	16A res.
Neutral	None	

Proscon PB°C

Common	3	Scale 2–27°C
Demand	1	H66 × W100 × D39
Satisfied	2	6A res. (2A)
Neutral	N	

Proscon PBF

As Proscon PBC but scaled in Farenheit.

Proscon R 1

Common	Yes	Scale 40–80°F
Demand	Yes	H128 × W73 × D61
Satisfied	Yes	20A res.
Neutral	None	

Terminals unmarked.

Randall R504D

Common	3	Scale 5–30°C
Demand	1	H84 × W84 × D45
Satisfied	2	16A res.
Neutral	N	

Randall R504N

Common	3	Scale 1–6°C
Demand	1	H84 × W84 × D45
Satisfied	2	16A res.
Neutral	N	

Randall R505D

With night set-back facility

Common	3	Scale 5–30°C
Demand	1	H84 × W84 × D45
Satisfied	2	16A res.
Neutral	N	

Randall R505N

Common	3	Scale 1–6°C
Demand	1	H84 × W84 × D45
Satisfied	2	16A res.
Neutral	N	

Randall RD 3

Common	1	Scale 5–30°C
Demand	2	H79 × W79 × D27
Satisfied	None	10A res. (4A)
Neutral	4	

Randall RSR/L

Common	3	Scale 13–24°C
Demand	1	H66 × W100 × D39
Satisfies	2	6A res. (2A)
Neutral	None	

Randall RSR/M

Common	3	Sale 13–24°C
Demand	1	H66 × W100 × D39
Satisfied	2	6A res. (2A)
Neutral	N	

Sangamo 925890

Common	1	Scale 5–35°C
Demand	3	H75 × W75 × D31
Satisfied	None	10A res.
Neutral	4	

Sangamo 925895

With on/off switch

Common	1	Scale 5–35°C
Demand	3	H75 × W75 × D31
Satisfied	None	10A res.
Neutral	4	

Sauter TSH 3

Common	1	Scale 4–30°C
Demand	2	H60 × W100 × D40
Satisfied	3	15A res. (1A)
Neutral	4	

Sauter TSH 57 (F004)

Common	2	Scale 5–30°C
Demand	3	H71 × W71 × D28
Satisfied	1	10A res. (4A)
Neutral	4	

Smiths RS

Common	3	Scale 6–28°C
Demand	2	H58 × W80 × D45
Satisfied	4	10A res.
Neutral	None	

Smiths ZV2521

Common	1	Scale 0–30°C
Demand	2	H72 × W72 × D36
Satisfied	None	16A res. (4A)
Neutral	4	

Smiths ZV2522

Common	1	Scale 0–30°C
Demand	2	H72 × W72 × D36
Satisfied	5	16A res. (4A)
Neutral	4	

Sopac TA 50 range

Common	2	Scale 45–80°F
Demand	3	H55 × W105 × D50
Satisfied	4	15A res.
Neutral	1	(if fitted)

Sopac TA 80

Common	3	Scale 6–28°C
Demand	2	H58 × W80 × D45
Satisfied	None	10A res.
Neutral	None	

Sopac TA 80Y

Common	3	Scale 6–28°C
Demand	2	H58 × W80 × D45
Satisfied	4	10A res.
Neutral	None	

Sopac TA 84

Common	3	Scale 6–28°C
Demand	None	H58 × W80 × D45
Satisfied	4	10A res.
Neutral	None	

Sopac TA 84Y

Common	3	Scale 6–28°C
Demand	2	H58 × W80 × D45
Satisfied	4	10A res.
Neutral	None	

Sopac TA 340

Common	1	Scale 6–30°C
Demand	3	H72 × W72 × D44
Satisfied	None	16A res. (2.5A)
Neutral	None	

Sopac TA 350

Common	1	Scale 6–30°C
Demand	3	H72 × W72 × D44
Satisfied	2	16A res. (2.5A)
Neutral	None	

Sopac TA 520

Common	1	Scale 6–30°C
Demand	3	H79 × W75 × D46
Satisfied	None	16A res. (2.5A)
Neutral	None	

Sopac TA 521

Common	1	Scale 6–30°C
Demand	3	H79 × W75 × D46
Satisfied	None	16A res. (2.5A)
Neutral	4	

Sopac TA 530

Common	1	Scale 6–30°C
Demand	3	H79 × W75 × D46
Satisfied	2	16A res. (2.5A)
Neutral	None	

Sunfine

Common	3	Scale 2–27°C
Demand	1	H66 × W100 × D39
Satisfied	2	6A res. (2A)
Neutral	N	

Sunvic TL range

T denotes with thermometer.

Sunvic TL 10 – 24V

Common	3	Scale 38–82°F
Demand	1	H67 × W103 × D65
Satisfied	None	1A res. (1A)
Neutral	None	

Sunvic TL 11 – 24V

Common	3	Scale 3–27°C
Demand	1	H67 × W103 × D65
Satisfied	None	1A res. (1A)
Neutral	None	

Sunvic TL 19

Common	3	Scale 3–27°C
Demand	1	H67 × W103 × D65
Satisfied	2	1A res. (1A)
Neutral	4	

Sunvic TL 25

Common	3	Scale 38–82°F
Demand	1	H67 × W103 × D65
Satisfied	2	1A res. (1A)
Neutral	4	

Sunvic TL 35

Common	3	Scale 38–82°F
Demand	1	H67 × W103 × D65
Satisfied	None	1A res. (1A)
Neutral	4	

Sunvic TL 39

Common	3	Scale 3–27°C
Demand	1	H67 × W103 × D65
Satisfied	None	1A res. (1A)
Neutral	4	

Sunvic TLM 2253

Common	3	Scale 3–27°C
Demand	1	H90 × W86 × D46
Satisfied	None	16A res.
Neutral	4	

Sunvic TLM 2257

Frost thermostat

Common	3	Scale −15–10°C
Demand	1	H90 × W86 × D46
Satisfied	None	16A res.
Neutral	None	

Sunvic TLM 2453

With tamper-proof cover

Common	3	Scale 3–27°C
Demand	1	H90 × W86 × D46
Satisfied	None	16A res.
Neutral	4	

Sunvic TLX 2222

Common	3	Scale 1–5°C
Demand	1	H90 × W86 × D46
Satisfied	None	6A res. (2.5A)
Neutral	4	

Sunvic TLX 2251 – 24V

Common	3	Scale 3–27°C
Demand	1	H90 × W86 × D46
Satisfied	None	1A res. (1A)
Neutral	None	

Sunvic TLX 2259

Common	3	Scale 3–27°C
Demand	1	H90 × W86 × D46
Satisfied	None	6A res. (2.5A)
Neutral	4	

Sunvic TLX 2356

Common	3	Scale 3–27°C
Demand	1	H90 × W86 × D46
Satisfied	2	2A res. (1A)
Neutral	4	

Sunvic TLX 2358

With tamper-proof cover

Common	3	Scale 3–27°C
Demand	1	H90 × W86 × D46
Satisfied	2	2A res. (1A)
Neutral	4	

Sunvic TLX 2360

Frost thermostat

Common	3	Scale 0–15°C
Demand	1	H90 × W86 × D46
Satisfied	None	6A res. (2.5A)
Neutral	4	

Sunvic TLX 3101 Electronic

Common	1	Scale 5–30°C
Demand	8	H86 × W86 × D37
Satisfied	None	2A
Neutral	2	

Sunvic TLX 5101 Electronic

Common	1	Scale 5–30°C
Demand	8	H86 × W86 × D37
Satisfied	None	10A (4A)
Neutral	2	

Sunvic TLX 5201 Electronic

Common	1	Scale 5–30°C
Demand	6	H86 × W86 × D37
Satisfied	8	10A (4A)
Neutral	2	

Sunvic TLX 5701 24V Electronic

Common	1	Scale 5–30°C
Demand	8	H86 × W86 × D37
Satisfied	None	10A (4A)
Neutral	2	

Sunvic TLX 7501

Common	3	Scale 5–35°C
Demand	2	H86 × W86 × D31
Satisfied	1	A 2(1)A
Neutral	None	Batteries 2 × AA

Sunvic TM 12

Frost thermostat

Common	3	Scale 8–52°F
Demand	1	H67 × W103 × D65
Satisfied	None	1A res.
Neutral	4	

Sunvic TM 16

Common	3	Scale 38–82°F
Demand	1	H67 × W103 × D65
Satisfied	None	1A res.
Neutral	4	

Sunvic TM 56

Common	3	Scale 3–27°C
Demand	1	H67 × W103 × D65
Satisfied	None	1A res.
Neutral	4	

Switchmaster plug-in

Common	3	Scale 5–30°C
Demand	2	H85 × W85 × D45
Satisfied	1	4A res. (1A)
Neutral	5	

Switchmaster SRT 1/2

Common	1	Scale 6–30°C
Demand	3	H72 × W72 × D44
Satisfied	2	16A res. (2.5A)
Neutral	None	

Switchmaster SRT 3

Specification as SRT 4, but with Homewarm cover with heating/hot water scale. Can be replaced with conventional thermostat if necessary.

Switchmaster SRT 4

Common	1	Scale 6–30°C
Demand	3	H79 × W75 × D46
Satisfied	2	16A res. (2.5A)
Neutral	None	

Teddington FEB

Common	3	Scale 5–30°C
Demand	2	H85 × W85 × D45
Satisfied	1	4A res. (1A)
Neutral	5	

Thorn security optima

Optimum start temperature control

Switch live	S/L	Scale 5–30°C
Permanent live	P/L	H84 × W84 × D30
Demand	On	6A res. (3A)
Satisfied	Off	
Neutral	N	

Permanent live is optional and provides a 1 hour heating boost facility.

Tower DT

Common	3	Scale 5–25°C
Demand	1	H66 × W100 × D39
Satisfied	2	6A res. (2A)
Neutral	N	

Tower RS

Common	1	Scale 5–35°C
Demand	3	H76 × W76 × D30
Satisfied	None	6A (2A)
Neutral	4	

Tower SS

Common	1	Scale 0–30°C
Demand	2	H70 × W70 × D35
Satisfied	None	16A res. (4A)
Neutral	4	

Tower ST

Common	3	Scale 5–25°C
Demand	1	H66 × W100 × D39
Satisfied	None	6A res. (2A)
Neutral	N	

Trac RS10

Common	1	Scale 5–30°C
Demand	2	H78 × W78 × D43
Satisfied	3	16A res. (2.5A)
Neutral	None	

Tristat

Produced mainly for the commercial market, it is used in conjunction with a passive infra-red sensor and enables temperature to be reduced automatically, when a room is unoccupied.

Unity

Common	L	Scale 35–80°F
Demand	H	H137 × W70 × D48
Satisfied	None	15A
Neutral	N	

Vaillant VRT 378

Common	3	Scale 5–30°C
Demand	4	H64 × W110 × D24
Satisfied	None	10A res.
Neutral	5	

Vaillant VRT 9090

Common	3	Scale 5–30°C
Demand	4	H60 × W112 × D35
Satisfied	None	10A res.
Neutral	5	

Vokera

Common	2	Scale 5–30°C
Demand	5	H80 × W80 × D35
Satisfied	None	10A res. (2.5A)
Neutral	6	
Link	1–4	

NOTE: Terminals run 1 3 2 4 5 6

Wickes RS

Common	6	Scale 5–30°C
Demand	2	H80 × W80 × D27
Satisfied	None	6A res. (2.5A)
Neutral	4	

Worcester Digistat CD

See ACL-Drayton Digistat RF1.

Wylex

Common	L-in	Scale 38–80°F
Demand	L-out	H122 × W90 × D50
Satisfied	None	1A ind.
Neutral	N	

6

Motorized valves and actuators

ACL Biflo	Mid-position
672 BRO 340	¾″ BSP
679 BRO 340	22mm
773 BRO 337	1″ BSP
	Heating port B, hot water port A

This valve requires SPDT room and cylinder thermostats. It can only utilize a simple time switch or a small group of electromechanical programmers, e.g. Sangamo F410, F9, Randall 105, Tower/ACL MP, Tower/ACL FP and Horstmann Gem (see Chapter 2). These can now be replaced with electronic programmers of the type where links usually need to be fitted live and switch commons but these links are not required for this system (see pages 62 and 248).

ACL Lifestyle	2-port zone	Spring return
679H 308	22mm	Auxiliary switch SPST 5A
779H 335	28mm	Auxiliary switch SPDT 5A
	Standard colour flex conductors.	

ACL Lifestyle	Diverter	
679H 314	22mm	Inlet centre port
779H 336	28mm	Port A open when energized (usually central heating)

Available with auxiliary switch if required. Standard colour flex conductors.

ACL Lifestyle	Mid-position	
679H 340	22mm	Inlet centre port, heating port A, hot water port B
779H 340	28mm	

Auxiliary switch rating 5A. Standard colour flex conductors.

ACL Motortrol	2-port zone	BSP	Spring return
631 B308	½″ BSP	Old	Auxiliary switch SPST 5A
691 B308	½″ BSP	New	Auxiliary switch SPST 5A

672 B308	¾″ BSP	Old	Auxiliary switch SPST 5A
679 B308	¾″ BSP	New	Auxiliary switch SPST 5A
773 B335	1″ BSP	Old	Auxiliary switch SPST 5A
779 B335	1″ BSP	New	Auxiliary switch SPDT 5A

24V, 110V and energize to close – available as special. Standard colour flex conductors.

ACL Motortrol **Diverter** **BSP**

691 B314	½″ BSP	Inlet port C, port B open when energized
672 B314	¾″ BSP	(usually central heating)
679 B314	¾″ BSP	Standard colour flex conductors.
773 B336	1″ BSP	

Barlo 2PV 1 **2-port zone** **Spring return**

As ACL 670 H308

Barlo 3PV 1 **Mid-position**

As ACL 679 H340

Danfoss ABV-VMT **2-port zone** **Thermohydraulic**

ABV-VMT 15/8	15mm pumped only systems
ABV-VMT 22/8	22mm pumped only systems
AVB-VMT 28/8	28mm pumped only systems
AVB-VMT 15/2	15mm gravity only systems
ABV-VMT 22/2	22mm gravity only systems
ABV-VMT 28/2	28mm gravity only systems

ABV is the actuator part of the valve and is available in both 24V and 40V. They do not have and auxiliary switch.

Danfoss ABV-VMV **Diverter** **Thermohydraulic**

ABV-VMV-15	½″
ABV-VMV-20	¾″
ABV-VMV-25	1″
ABV-VMV-32	1¼″
ABV-VMV-40	1½″

ABV is the actuator part of the valve and is available in both 24V and 240V. They do not have an auxiliary switch. The VMV must always be installed as a mixing valve (two inlet ports) according to the flow direction arrows cast into the valve body. The VMV closes across main ports A–AB on rising spindle travel.

Danfoss DMV-2C **2-port zone** **Spring return**

22mm	Auxiliary switch SPST 3A (2A)
28mm	Auxiliary switch SPDT 3A (2A)

24V version available. Standard colour flex conductors.

Danfoss DMV-21 — **2-port zone** — **Spring return**

1″ BSP

Auxiliary switch SPDT 3A (2A)
Wiring as DMV-2C 28mm. Standard colour flex conductors.

Danfoss DMV-3D — **Diverter**

22mm

Inlet port AB
Port A open when energized (usually central heating)
Standard colour flex conductors

Danfoss DMV-3M — **Mid-position**

22mm

Inlet port AB. Heating port A. Hot water port B
Auxiliary switch rating 3A (2A). Standard colour flex conductors.

Danfoss HP 2 — **2-port zone**

As Randall HP 2

Danfoss HS 3 — **Mid-position**

As Randall HS 3

Drayton TA/M2 — **2-port zone actuator – motor open/close**

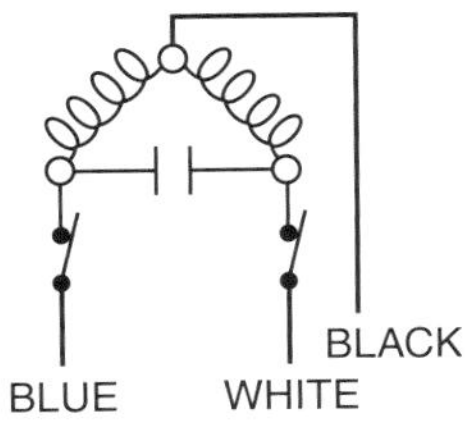

Figure 6.1 *Drayton TA/M2*

Fits to TA/VA range of valve bodies – see after TA/M5
Energize WHITE for clockwise rotation of actuator (Port 2 shut)
Energize BLUE for anti-clockwise rotation of actuator (Port 3 shut)
Neutral BLACK
No auxiliary switch
See also page 249

Drayton TA/M2A — **2-port zone actuator – motor open/close**

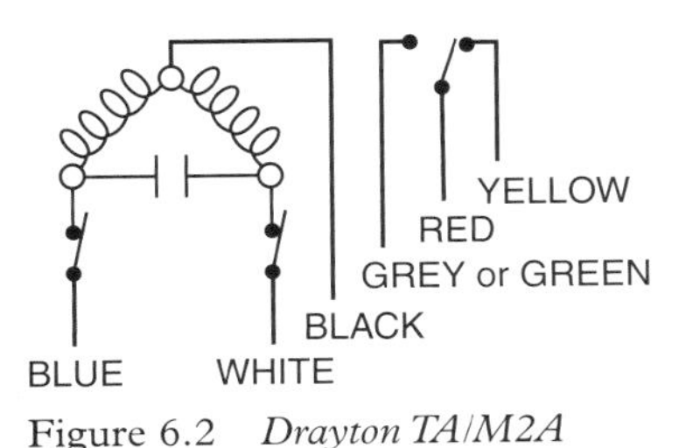

Figure 6.2 *Drayton TA/M2A*

Fits to TA/VA range of valve bodies – see after TA/M5
Energize WHITE for clockwise rotation of actuator (Port 2 shut)
Energize BLUE for anti-clockwise rotation of actuator (Port 3 shut)
Neutral BLACK
Auxiliary switch SPDT 3A
Valve clockwise RED + YELLOW made
Valve anti-clockwise RED + GREY made
See also page 251

Drayton TA/M4

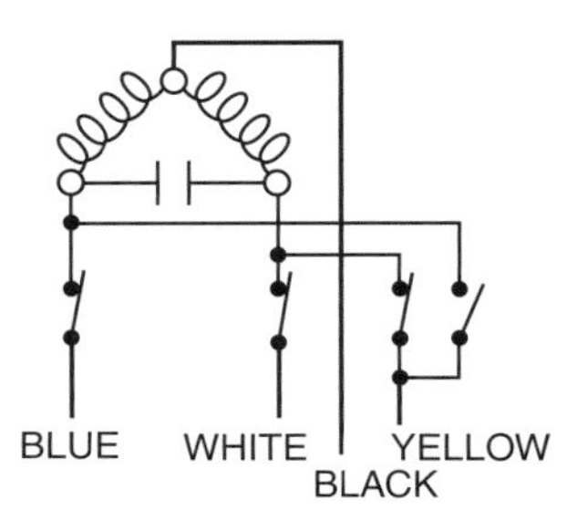

Figure 6.3 *Drayton TA/M4*

Mid-position actuator

Fits to TA/VA range of valve bodies – see after TA/M5.
Usually used in conjunction with an RB1 or RB2 relay box for boiler switching – see pages 101
Energize WHITE for clockwise rotation of valve (Port 2 shut)
Energize BLUE for anti-clockwise rotation of valve (Port 3 shut)
Energize YELLOW for mid-position
Neutral BLACK
See also page 250

Drayton TA/M5

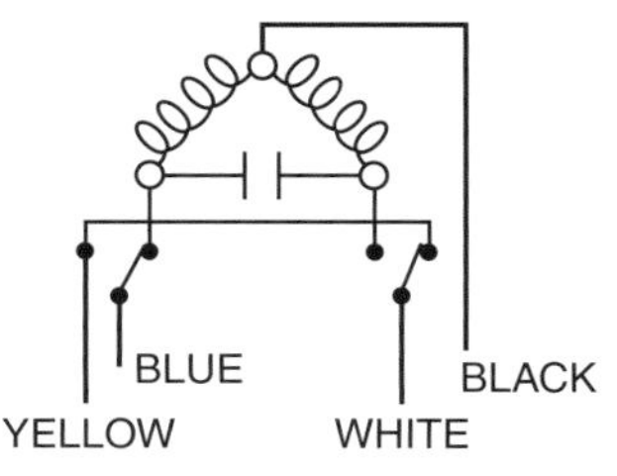

Figure 6.4 *Drayton TA/M5*

Diverter actuator

Fits to TA/VA range of valve bodies – see over
Energize WHITE for clockwise rotation of valve (Port 2 shut)
Energize BLUE for anti-clockwise rotation of valve (Port 3 shut)
Auxiliary switch SPDT 3A
Valve clockwise YELLOW + WHITE made
Valve anti-clockwise YELLOW + BLUE made

Drayton TA/VA

Valve bodies

TA/V1 ½″ 15 mm 2-port
TA/V2 ½″ 15 mm 3 port inlet port 1
TA/V4* ¾″ BSP 3 port, ports 2 and 3 reversible
TA/V6* 1″ BSP 3 port

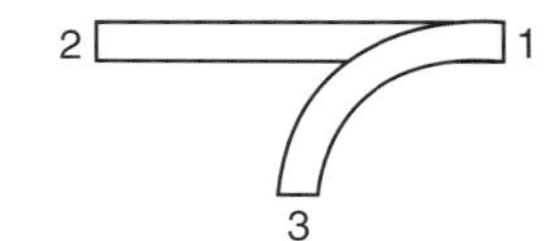

*Can be converted to 2-port by plugging third port.

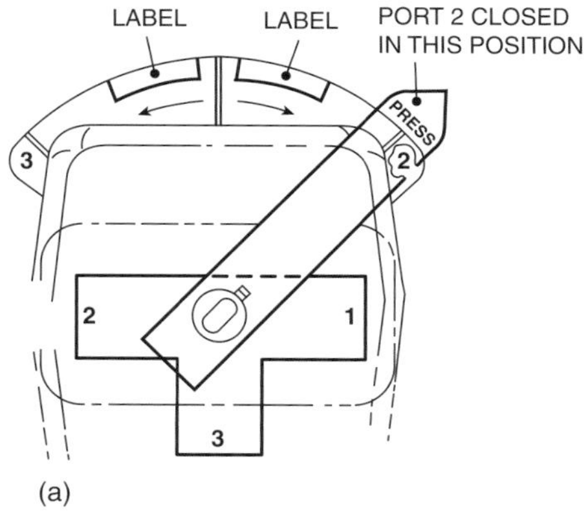

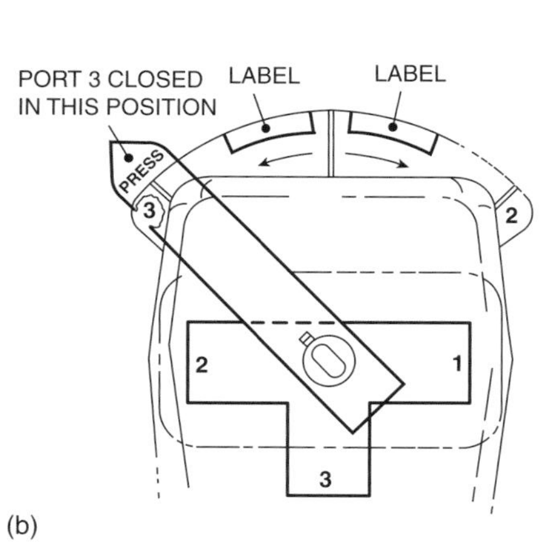

Figure 6.5 *Drayton TA/VA*

Drayton RB1 and RB2 relay boxes (usually used with TA/M4 actuator)

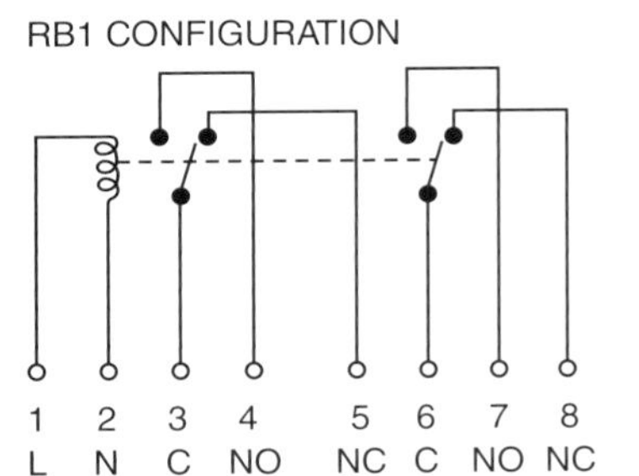

Figure 6.6 *Drayton RB1*

Double pole, double throw relay with no internal links.H95 × W111 × D71
Contact rating 6A (1.5A)
See also page 250

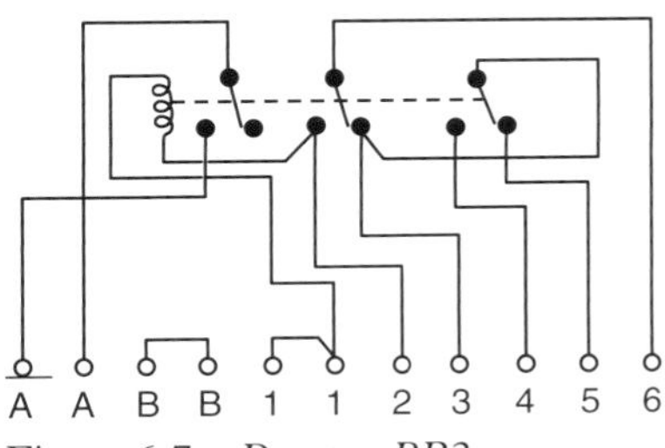

Figure 6.7 *Drayton RB2*

Double pole, double throw relay with printed circuit internal links.H95 × W111 × D71
Contact rating 6A (2A)
See also page 250

Drayton ZVA 22 **2-port zone** **Spring return**

22mm

Auxiliary switch SPST 3A
Standard colour flex conductors.

Drayton ZVA 28 **2-port zone** **Spring return**

28mm

Auxiliary switch SPDT 3A
Standard colour flex conductors.

Honeywell V 2057A

24V version of V 6057A

Honeywell V 4043 **2-port zone** **Spring return**

B1257	22mm	De-energized OPEN	Auxiliary switch none – see page 243
B1265	28mm	De-energized OPEN	Auxiliary switch none – see page 243
C1156	½″ BSP	De-energized SHUT	Auxiliary switch none
H1056	22mm	De-energized SHUT	Auxiliary switch SPST 2.2A
H1106	28mm	De-energized SHUT	Auxiliary switch SPDT 2.2A
H1007	¾″ BSP	De-energized SHUT	Auxiliary switch SPST 2.2A
H1080	1″ BSP	De-energized SHUT	Auxiliary switch SPDT 2.2A

Standard colour flex conductors.

Honeywell **V 4044C** **Diverter**

1288 — 22mm — Inlet Port AB
1569 — 28mm
1098 — ¾″ BSP — Port A open when energized
1494 — 1″ BSP — (usually central heating)

Standard colour flex conductors.

Honeywell V 4073A **Mid-position 5 wire**

1039 — 22mm — Inlet port AB, heating port A, hot water port B
1088 — 28mm
1054 — ¾″ BSP — Auxiliary switch rating 2.2A
1062 — 1″ BSP — Standard colour flex conductors.

Honeywell V 4073 **Mid-position 6 wire**

Inlet port AB
Heating port A
Hot water port B

Identifiable by the external relay fitted to the valve motor cover. If replacing with V 3073 5-wire connect colour for colour (disregard brown wire connection) and change over wires on cylinder thermostat common and demand. See also pages ?

Honeywell V 6057A **2-port zone** **Motor open/close**

Fitted to a V 5057A valve body, ¾″ or 1″
Fitted with or without SPDT
Auxiliary switch 5A (2A)
Terminal 1 Neutral
Terminal 2 Close valve
Terminal 3 Open valve

Auxiliary switch contacts if fitted:
Terminal 4 Common
Terminal 5 Made when valve shut
Terminal 6 Made when valve open

Honeywell V 8043 **2-port zone**
V 8044 **Diverter**
V 8073 **Mid-position**

The above valves were 24V versions of the 240V range. Should any of the above need replacing, then it can be done so with the 240V equivalent and a 24V motor which is available as a spare.

Hortsmann **2-port zone** **Spring return**

H-2-22-Z — 22mm — Auxiliary switch SPST
H-2-28-Z — 28mm — Auxiliary switch SPDT

Standard colour flex conductors.

Hortsmann **Mid-position**

H-3-22-M — 22mm

Standard colour flex conductors.

Horstmann **2-port zone** **Spring return**

Z222XL 22mm Auxiliary switch SPST 3A
Standard colour flex conductors.

Horstmann **2-port zone** **Spring return**

Z228XL 28mm Auxiliary switch SPDT 3A
Standard colour flex conductors.

Horstmann **Mid-position 4 wire**

Z322XL 22mm

Inlet port AB, heating port A, hot water port B
Auxiliary switch rating 3A
Standard colour flex conductors.

Landis & Gyr Da-V322 **Diverter**

22mm

Inlet port AB. Port A open when energized (usually central heating)
Standard colour flex conductors.

Landis & Gyr MA-V3 **Mid-position**

MA-V322 22mm
MA-V328 28mm

Inlet port AB, heating port A, hot water port B
Auxiliary switch rating 3A
Standard colour flex conductors.

Landis & Gyr SK2/LL **2-port zone** **Spring return**

LL4402 15mm valve body
LL4453 22mm valve body
LL4501 28mm valve body
SK2 Actuator

Auxiliary switch SPST 5A (5). Standard colour flex conductors.

Landis & Gyr SK3/2701 **Mid-position**

LT2701 22mm valve body Inlet port AB, heating port A
SK3 Actuator Hot water port B

Auxiliary switch rating 5A (5). Standard colour flex conductors.

Landis & Gyr STC 4 × 7¾″ Mid-position

This valve was used only in conjunction with a relay box. See page 253

Landis & Gyr ZA V2	**2-port zone**	**Spring return**
ZA-V215	15mm	Auxiliary switch SPST 3A
ZA-V222	22mm	Auxiliary switch SPST 3A
ZA-V228	28mm	Auxiliary switch SPST 3A
		Standard colour flex conductors.

Myson MSV222	**2-port zone**	**Spring return**
22mm	Auxiliary switch rating 3A Standard colour flex conductors.	

Myson MSV228	**2-port zone**	**Spring return**
28mm	Auxiliary switch rating 3A Standard colour flex conductors.	

Myson MSV322	**Mid-position**
22mm	Auxiliary switch rating 3A Inlet port AB, heating port A, hot water port B Standard colour flex conductors.

Potterton MSV322	**Mid-position**
28mm	Auxiliary switch rating 3A Inlet port AB, heating port A, hot water port B Standard colour flex conductors.

Potterton PMV2	**2-port zone**	**Spring return**
22mm	As Landis & Gyr SK2	

Potterton PMV3	**Mid-position**
22mm	As Landis & Gyr SK3

Randall HP **2-port zone** **Spring return**

As Switchmaster VM4 auto zone valve

Randall HP 2 **2-port zone** **Spring return**

HPA 2	**Actuator**	Auxiliary switch SPST
HPA 2C	**Actuator**	Auxiliary switch SPDT

Standard colour flex conductors.

HS 3	**Diverter**
HSA 3D	**Actuator**

Standard colour flex conductors.

Randall HS **Mid-position**

As Switchmaster VM1 Midi

Randall HS 3 **Mid-position**

HSA 3 **Actuator**
Standard colour flex conductors.

Siemens See Landis & Gyr

Smiths Centroller O-C/DV 1-¾″ **Diverter** **Thermohydraulic**

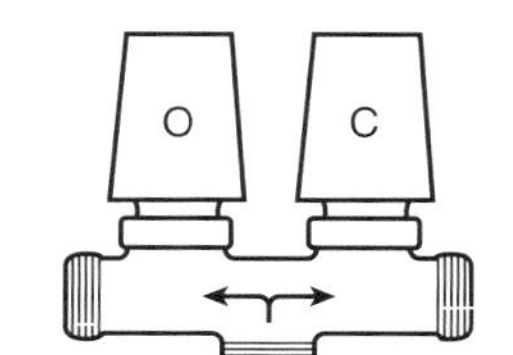

Figure 6.8

This diverter valve is made up of two separate Actuators, one code O (energized to open) and one code C (energized to close). Actuator O should be in the central heating branch of the valve body.

Smiths Centroller OS/V22 **2-port zone** **Thermohydraulic**

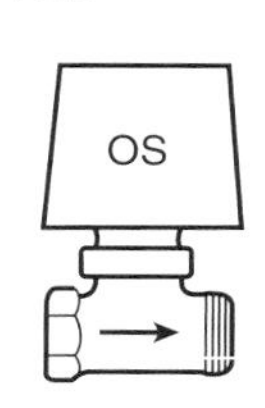

Figure 6.9

Consisting of a V22 or V¾″ valve body and an OS energized to open actuator with SPST auxiliary switch. If no auxiliary switch required, then a Code O actuator can be used.

Smiths Centroller OS/DV 1-¾″	**Mid-position**	**Thermohydraulic**
		Consisting of a DV 1-¾″ valve body and two OS energized to open actuators

Figure 6.10

Smiths Centroller O-C-OS	**Actuators**	**Thermohydraulic**

Code O energize to open
Code C energize to close
Code OS energize to open with auxiliary switch

These actuators work by a heated wax cylinder, operating a plunger to open or close the appropriate port and operate an end switch if fitted. They are, therefore, wired as a spring return type actuator, but take considerably longer to operate. They were not pre-wired.

Sopac ZV	**2-port zone**	**Spring return**
ZV-15-2	½″ BSP	Auxiliary switch to all models
ZV-20-2	¾″ BSP	SPST switch rating 3A
ZV-20-2-B	22mm	Energize to close available with or without auxiliary switches
ZV-25-2-B	28mm	

Sopac ZV	**Diverter**	
ZV-20	¾″ BSP	Inlet port AB
ZV-25	1″ BSP	Port A open when energized (usually central heating)
ZV-20-EB	22mm	
ZV-25-B	28mm	

Sopac ZV	**Mid-position**	
ZV-20-EB-MID	22mm	Inlet port AB, heating port A, hot water port B
ZV-25-B-MID		Auxiliary switch rating 3A
		Standard colour flex conductors.

Sunvic DM 3601 **Mid-position**
DM 3551

For use only with the Duoval RJ relay box.
DM3651 has a 5-way plug for connection into the Duo-plug RJ relay box.

Sunvic DM 4601 **Mid-position**
DM 4651

For use only with the Duoval RJ relay box.
DM4651 has a 5-way plug for connection into the Duo-plug RJ relay box.
The DM 36 and 46 Duoval and Duo-plug models are interchangeable and differ only in styling. They can also be connected into any of the three relay boxes, RJ 1801, RJ 2801 and RJ 2802 or the Duo-plug relay box RJ 2852 using the appropriate wiring diagram. See pages 255 and 256

Sunvic DT and EDT **Valve bodies for use with SD actuators**

DT 1601	¾″ BSP
DT 1701	22mm
DT 1801	Has a 28mm connection to inlet and port A and a 22mm connection to port B
DT 2601	¾″ BSP
EDT 1702	22mm
EDT 2702	22mm

Sunvic Duoflow – RJ 1801 relay box

1	2	3	4	5	6	7	8	9	10	11	12	13	14	15	16
L	N	E	L	N	YEL	ORA	WHI	BLU	N	E	DEM	COM	DEM	SAT	COM
MAINS			BOILER PUMP		ACTUATOR						ROOM	STAT	CYLINDERSTAT		

If boiler has pump overrun, pump should be wired as boiler instructions. See also page 255

Sunvic Duoflow – RJ 2801 relay box

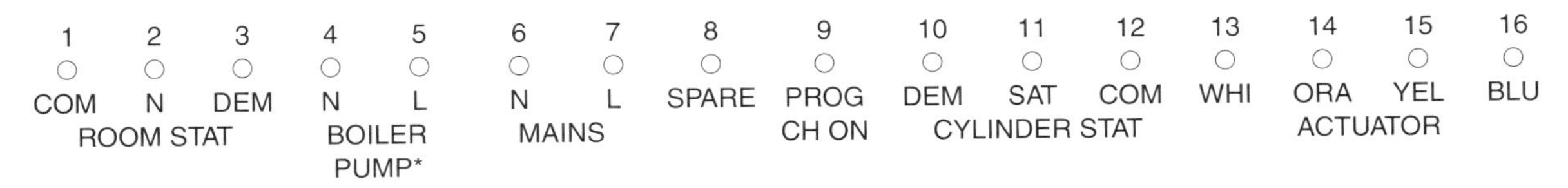

If programmer has HW OFF connection wire to terminal 11.
*If boiler has pump overrun, pump should be wired as boiler instructions.
See also page 255

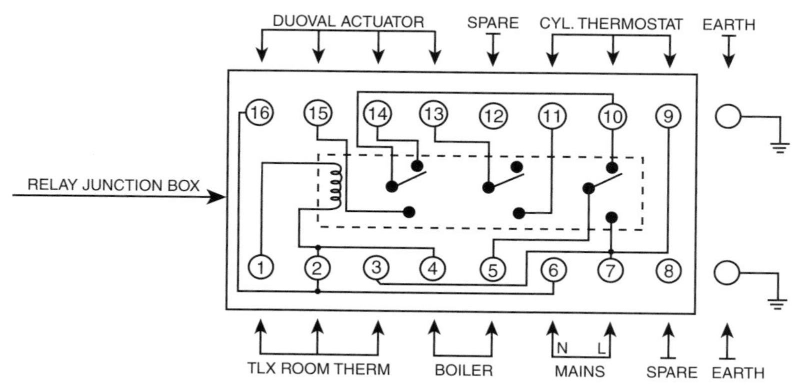

Figure 6.11 *Sunvic Duoflow RJ 2801 relay box, internal diagram*

Sunvic Duoflow RJ 2802 Relay box
RJ 2852 Relay box (plug-in)

1	2	3	4	5	6	7	8	9	10	11	12	13	14	15	16	17	18	19	20
DEM	N	COM	N	L	N	PROG	L	COM	DEM	SAT	SPARE	WHI	ORA	YEL	BLU				PRO
ROOM STAT			BOILER PUMP*			CH ON MAINS		CYLINDER STAT				ACTUATOR							HW ON

If programmer has HW OFF connection wire to terminal 18.
*If boiler has pump overrun, pump should be wired as boiler instructions.
See also page 256

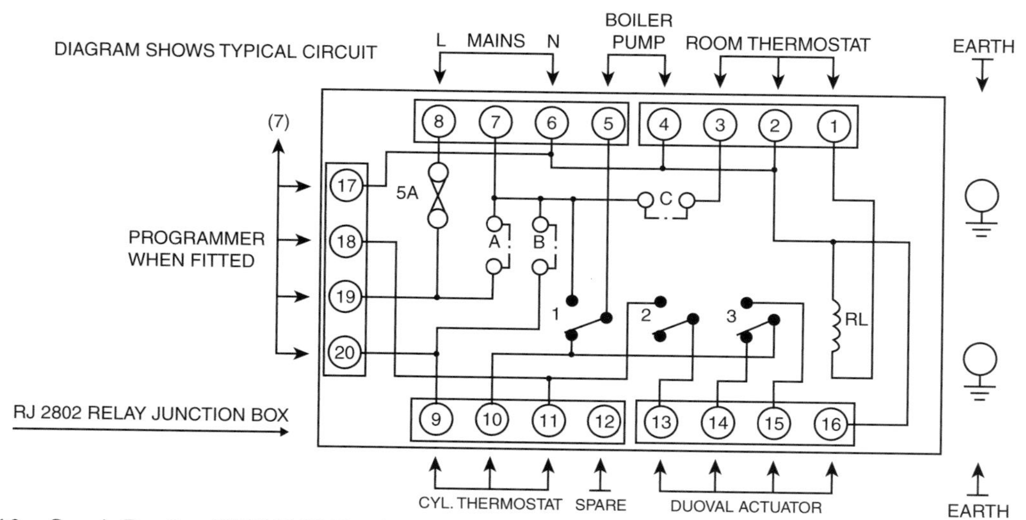

Figure 6.12 *Sunvic Duoflow RJ 2802/2852, relay box, internal diagram*

Sunvic SD 1601 **Diverter**
SD 1626

Combines with the DT or EDT valve body to make a priority valve in the Uniflow system. The SD 1626 differs in that it has a manual lever.

Sunvic SD 1701 **Mid-position**
SD 1726

Combines with the DT or EDT valve body to make a mid-position valve in the Unishare system. The SD 1726 differs in that it has a manual lever.
Standard flex covers.

Complete Valve **SDV 1211** 22mm

Sunvic SD 1752 **Mid-position**

Combines with the EDT 1702 valve body to make a mid-position valve in the Clockbox II Unishare system. The actuator is fitted with a 4-way plug.

Sunvic 2601 Replacement actuator for SD 1601.

Sunvic SD 2701 Replacement actuator for SD 1701.

Complete valve **SDV 2211** 22mm

Sunvic SM2, SM2201 **2-port actuator now SM 4201**

Sunvic SM5, SM 2203 **2-port actuator now SM 4203**

Sunvic SML **2-port zone** **Motor open/close**

A range of 2-port motorized valves comprising of a valve body type ML and an actuator type SM to make the Minival series. The actuator range was initially SM 3201–5 and then restyled to become SM 4201/3/5 and later just SM 5201/3. The wiring principle has remained the same, although the colours were altered slightly when blue instead of black became neutral. All models fit the ML valve body.

ML 3401	½″ BSP	Valve body
ML 3402	15mm	Valve body
ML 3451	¾″ BSP	Valve body
ML 3453	22mm	Valve body
ML 3501	1″ BSP	Valve body
EML 3501	28mm	Valve body

SM 3201/4201/5201	Auxiliary switch None	18″ lead
SM 3202	Auxiliary switch None	72″ lead
SM 3203/4203	Auxiliary switch Yes	36″ lead
SM 3204/5203	Auxiliary switch Yes	17″ lead
SM 3205/4205	Auxiliary switch Yes	18″ lead
	Auxiliary switch rating 3A (1.3A)	

On the SM 3203/4203 and 3204 the auxiliary switch provides a live supply on the orange wire when the valve is open.

On the SM 3205/4205 the auxiliary switch provides a live supply on the orange wire when the valve is open and a live supply on the pink wire when the valve is closed.

The EML valve body differs from the ML range in having external compression fittings.

Colour code	*Old*	*New*
Neutral	Black	Blue
Motor open	Yellow	Yellow
Motor close	Blue	White
Aux. sw. (if fitted) open	Orange	Orange
closed	Pink	Pink

Sunvic SZ **2-port zone** **Spring return**

A range of spring return actuators that fit to the ML and EML range of valve bodies, as above, to make the Unival series.

MK1 MK2		
SZ 1201/2201	Auxiliary switch None	De-energized SHUT
SZ 1226/2226	As SZ 1201, with manual lever	
SZ 151/2251	Auxiliary switch None	De-energized OPEN
SZ 1301/2301	Auxiliary switch SPST	De-energized SHUT
SZ 1302/2302	Auxiliary switch Yes Designed for use on gravity hot water systems. See also page 237	De-energized SHUT
SZ 1326/2326	As SZ 1301, with manual lever	
SZ 1327/2327	As SZ 1302, with manual lever	
SZ 1351/2351	Auxiliary switch SPST	De-energized OPEN
SZ 2301 F	Functionally the same as the SZ 2301 but has a superior quality motor for use in high ambient temperature locations such as within boiler casings.	
SZV 2212	Complete unit	
SZV 2218	Complete unit	
SZV 2228	Complete unit	

Switchmaster VM1 **Mid-position**

Consisting of type VB1 valve body (22mm) and a type VA1 actuator. Known as a MIDI valve, it is blue in colour. Actuator can be turned with a screwdriver for manual operation if necessary. See also page 257

Inter – centre T
Central heating port – right hand with T at bottom
Hot water port – left hand with T at bottom
Auxiliary switch rating 3A

Switchmaster VM2 **2-port zone** **Motor open/close**

Consisting of type VB2 valve body (22mm) and a type VA2 actuator, which is blue in colour. Actuator can be turned with a screwdriver for manual operation if necessary.

Auxiliary switch rating 3A

Switchmaster VM3 **Mid-position**

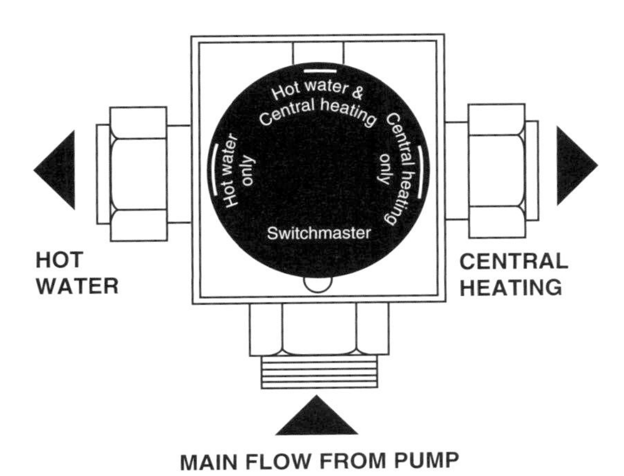

Figure 6.13

This is a manual version of the VM1 on the VB1 valve body.

Switchmaster VM4 **2-port zone** **See description**

Consisting of type VB4 valve body (22mm) and a type VA4 actuator, which is brown in colour and labelled AUTOZONE. Actuator can be turned with a screwdriver for manual operation of necessary. Although the actuator is of the motor open/close type, it is wired as and has the same colour coding as a spring return actuator, although it is necessary to ensure that the grey of the auxiliary switch is connected to live, and the orange is connected to boiler and pump.

Switchmaster VM5 **Diverter**

Consisting of type VB5 valve body (22mm) and a type VS5 actuator, which is brown in colour. The valve is designed specifically for the HOMEWARM system and is wired internally to give hot water priority when the room thermostat is satisfied. When the room thermostat is calling for heat, then 90% of boiler output goes to the heating circuit and the remaining 10% to the hot water circuit. A 1m lead is supplied, exposing four colours. However, if the cable is shortened, an orange wire will be exposed and can be disregarded.

Same port format as VM1. See also page 252

Tower MP 3	**Mid-position**
MP 332C	22mm
MP 3-1″B	1″ BSP available special order only
MP 3-28C	28mm available special order only

Auxiliary switch rating 3A. Standard colour flex conductors. Inlet port AB, heating port A, hot water port B

Tower MV 2	**2-port zone**	**Spring return**
MV2-22C	22mm	Auxiliary switch SPST 3A
MV2-1″B	1″	Auxiliary switch SPDT 3A
MV2-28C	28mm	Auxiliary switch SPST 3A
	Standard colour flex conductors.	

Tower MV3-22C	**Diverter**
22mm	Inlet port AB, heating port A, hot water port B Standard colour flex conductors.

Tower/ACL

Many of the earlier Motortrol range of motorized valves were under the TOWER and ACL name. Refer to ACL.

Wickes	**2-port zone**
As Landis & Gyr SK 2	

Wickes	**Mid-position**
As Landis & Gyr SK 3	

7

Boilers – general

Electricity

The use of electricity as a fuel for wet system central heating systems is limited to only a few manufacturers even though electricity as a fuel gives the advantage of being able to site the boiler virtually anywhere as no flue or storage of fuel is required. Also some are able to utilize off-peak electricity. They do however require a substantial electrical supply, e.g. 50A in the case of 12 kW versions, and this may be a governing factor when considering options.

Gas

Gas boilers come in all shapes, sizes and formats, and as such have the advantage of being able to be utilized for virtually any application. They can be floor, wall or hearth mounted and heat exchanger material could be cast iron, copper, aluminium, stainless steel, or even a mixture of metals. Even where there is no mains gas supply, there is usually an LPG model that could be used, although the siting of the LPG storage vessel could be problematical.

Gas boilers can be used in any wet heating system and, of course, gas fuels many warm air systems that may incorporate domestic hot water circulators. In addition to a basic boiler, other models available include: boilers with integral sealed system equipment such as pressure vessel, gauge, etc.; combination boilers which offer instant hot water and remove the need for a hot water storage cylinder; condensing boilers which recycle the hot flue gases to produced over 90 per cent efficiency; and system boilers which include pump, motorized valve and possibly programmer within its casing, thereby usually making installation easier.

As regards electrical control wiring, basic boilers comes in two types – those requiring just a switched live, neutral and earth connections often via an integral 3-pin plug, or the type of boiler which has a pump over-run facility. This device is a pre-set thermostat that measures the temperature of the heat exchanger. When in operation the over-run facility enables the system pump to continue running for a short period, possibly up to 20 minutes after the boiler itself has shut down thus dissipating the residual heat from the heat exchanger. This method is generally employed on wall-hung, low water content boilers that use non-cast iron heat exchangers, although some manufacturers of boilers using cast iron heat exchangers also include this over-run facility.

In addition to the above, some boilers may require additional wiring if they incorporate a programmer or timer, or in the case of back boiler units, a permanent live supply may be needed so that the live fuel effect light bulbs can be utilized when the boiler is off. It is essential that this live supply be taken from the same source as that for the boiler itself.

One type of gas boiler not mentioned in this book is those that use a pressure jet gas burner. This is because it is used on boilers intended for commercial and industrial premises. However, they can be linked into the external controls described here, in the same way as those boilers included.

Oil

Like gas, oil boilers come in a wide range of types suitable for floor, wall and hearth mounting. Oil also offers the basic boiler, combination and systems boilers. There are two type of oil boiler – wall flame and pressure jet. Wall flame boilers are the less popular and operate by oil being pumped into a large drum where it is electrically ignited, thereby heating the heat exchanger. The pressure jet fires oil into the burner under pressure and is also ignited electrically. These are the most popular type of oil boiler. Both types employ safety lock-out features to prevent oil being discharged without firing. Oil boilers need to be finely tuned in much the same way as a car engine so that it operates reliably and with maximum efficiency. It is essential that the correct grade of oil is used for each type of burner.

The method of wiring to oil boilers varies in that the supply wiring could go into the wall flame control box, the pressure jet control box or into a boiler terminal strip probably located at the top of the casing near the boiler thermostat. Wiring details are given for popular models but don't be surprised if some look as if they are not included. This may be because they employ a control box as listed and wiring details are given for those instead. Manufacturers and models where this may be the case include: Delheat, Heatrae Aggressor, McFarlane, Perrymatic and Potterton.

Solid fuel

By their very nature solid-fuel boilers only come as floor standing models, although electrical controls can be incorporated into boilers fitted into fireplaces as room heaters. These latter models can have a heat exchanger for supplying hot water to the domestic cylinder or heating circuit using gravity pumped or circulation. The Honeywell Y605B Link Fuel Plan described in the ancillary equipment section (Chapter 11) may be of use to anyone wishing to install or adapt such a system. Features of a conventional solid-fuel boiler may include a fan to aid combustion, or a safety switch which will shut down the boiler in the event of chimney blockage.

Because the temperature of the fuel cannot be instantly reduced or shut down as with gas or oil, solid-fuel boilers cannot be employed in conventional fully pumped heating systems. However, by using a combination of normally-open and normally-closed spring return motorized valves it is possible to utilize a solid-fuel boiler in such a system. The Honeywell Y605A Panel Solid Fuel Timed Sundial Plan is a fully assembled, pre-plumbed and pre-wired control set and is designed specifically for use with a solid-fuel boiler on a fully pumped system.

A feature of the Trianco TRG range of boilers is the use of an Economy Thermostat which is preset at a low temperature (57°C) for night operation and any day period when central heating is not required. A double circuit programmer is used to control the boiler. When hot water is programmed, the boiler operates on the Economy Thermostat. When central heating is selected, the boiler is controlled by the higher thermostat setting and the pump also starts (see Figure 7.1).

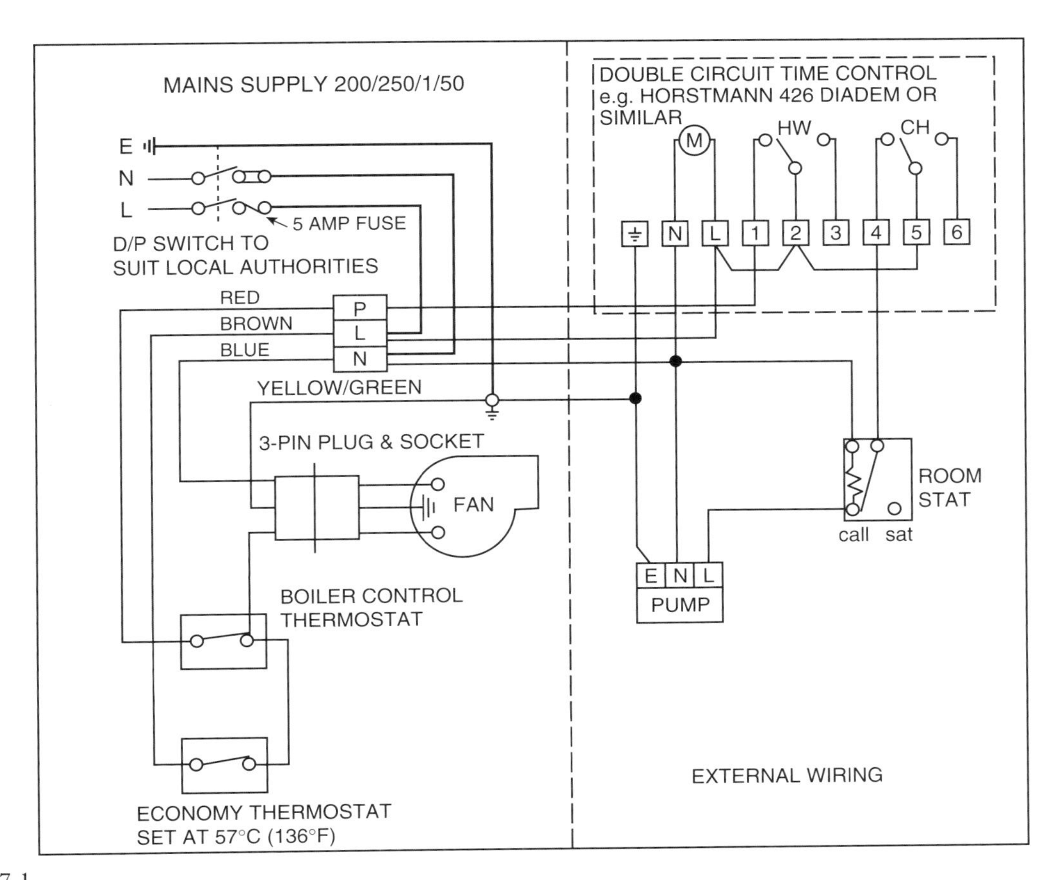
MAINS SUPPLY 200/250/1/50
E
N
L
5 AMP FUSE
D/P SWITCH TO
SUIT LOCAL AUTHORITIES
RED
BROWN
BLUE
P
L
N
YELLOW/GREEN
3-PIN PLUG & SOCKET
FAN
BOILER CONTROL
THERMOSTAT
ECONOMY THERMOSTAT
SET AT 57°C (136°F)
DOUBLE CIRCUIT TIME CONTROL
e.g. HORSTMANN 426 DIADEM OR
SIMILAR
M
HW
CH
N
L
1
2
3
4
5
6
ROOM
STAT
call sat
E N L
PUMP
EXTERNAL WIRING

Figure 7.1

8

Boilers – electric

Electroheat Amptec

The boiler requires two supplies; a 6A supply for the control circuit and an additional supply for the heating elements depending on model selected. In all cases an RCD rated at 30mA must be fitted.

Control Circuit						Supply		
T	N	R	E	N	L	N	E	L
○	○	○	○	○	○	○	○	○
	N	Live | in		Pump		N	E	L

Fully pumped systems only
Suitable for use on sealed systems (with 2-port safety motorized valve)
Efficiency 99.8%

It may be necessary to fit a capacitor between N and R to prevent the boiler continuing to run due to leakage current from some motorized valves.

Gledhill Electromate 2000

The boiler provides mains pressure hot water and conventional central heating by radiators by using off-peak electricity to heat the thermal store. When the store is depleted the controller starts the recharge cycle giving priority to times when off-peak electricity is available. It therefore requires standard tariff and off-peak supplies.

A programmable room thermostat, supplied with the unit, best serves control of heating.

26	25	27	P28
○	○	○	○
Room Stat Common		E	Room Stat Demand

Fully pumped systems only
Suitable for use on sealed systems
Heat exchanger material

Myson Maxton

This boiler dates back to the 1980s but as they could still form part of a system upgrade, it is included. An off-peak supply is desirable but not essential.

N	6	8
○	○	○
N	Control Live	Mains Live

Fully pumped systems only
Heat exchanger material, copper
Efficiency 99%

Trianco Aztec

The boiler requires two supplies; a 6A supply for the control circuit and an additional supply for the heating elements depending on model selected. In all cases an RCD rated at 30mA must be fitted.

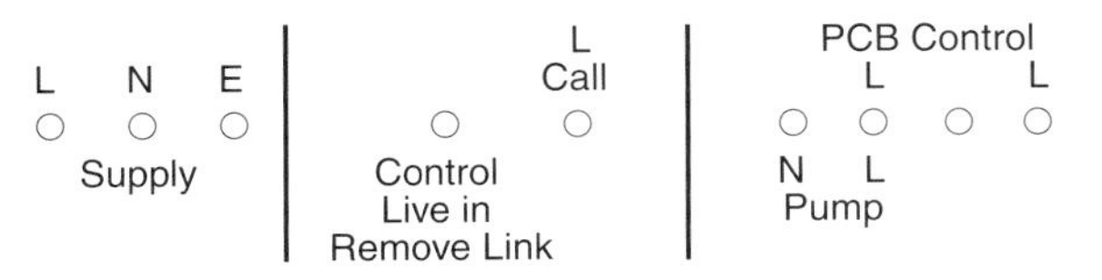

Fully pumped systems only
Suitable for use on sealed systems
Heat exchanger material, copper
Efficiency 99.8%

9

Boilers – gas

Alpha CB24/28 combination

Wall mounted

A room thermostat suitable for mains voltage should be connected to terminals 1–2 after removing link.

Fully pumped systems only
Suitable for sealed systems
Heat exchanger material, copper
Integral frost protection
SEDBUK rating D

Alpha SY9-24 System

Wall mounted

F2A	L	N	E	1	2
○	○	○	○	○	○
Fuse	L	N	E	Switch Live	Live Out

Remove link 1–2 if fitting external controls

Fully pumped systems only
Suitable for sealed systems
Heat exchanger material, copper
Integral frost protection
SEDBUK rating D

Alpha SY 24 System

Wall mounted

A room thermostat suitable for mains voltage should be connected to terminals 1–2 after removing link.

Fully pumped systems only
Suitable for sealed systems
Heat exchanger material – Copper
Integral frost protection
SEDBUK Rating D

Ariston 20 MFS combination

Wall mounted

Boiler is supplied with a 3-core mains lead.
External controls such as a time clock with voltage free contacts, or room thermostat suitable for use on 240V, should be connected as follows to the 12-way terminal strip:

TT	TR	TC	N	L
○	○	○ —	○	○
OUT	N	IN	Mains	

Suitable for sealed systems
Heat exchanger material, copper

Ariston DIA 20 MFFICE combination

Wall mounted

Boiler is supplied with a 3-core mains lead.
A room thermostat suitable for use on low voltage and/or can be connected into terminals after removing either of the two brown links from the green connector on the PCB board (an external timer with voltage free terminals).

Suitable for sealed systems
Heat exchanger material, copper
SEDBUK rating D

Ariston Eco Genus 24 condensing

Wall mounted

Connect a room thermostat to TA on terminal strip after removing link.

Fully pumped systems only
Suitable for sealed systems only
Heat exchanger material, aluminium
Integral frost protection
SEDBUK rating B

Ariston Eco Genus 24 system condensing

Wall mounted

Connect a room thermostat to TA on terminal strip after removing link.

Fully pumped systems only
Suitable for sealed systems only
Heat exchanger material, aluminium
Integral frost protection
SEDBUK rating A

Ariston Eurocombi combination

Wall mounted

Connect a room thermostat to control strip after removing link.

Fully pumped systems only
Suitable for sealed systems only
Heat exchanger material, copper
Integral frost protection
SEDBUK rating C

Ariston Genus 23, 27, 30 combination

Wall mounted

Connect a room thermostat to control strip after removing link.

Fully pumped systems only
Suitable for sealed systems
Heat exchanger material, copper
Integral frost protection
SEDBUK rating D

Ariston Genus 27 system

Wall mounted

Connect a room thermostat to control strip after removing link.

Fully pumped systems only
Suitable for sealed systems
Heat exchanger material, copper
Integral frost protection
SEDBUK rating D

Ariston Micro system

Wall mounted

L	N	1	2	3	4	5	6
○	○	○	○	○	○	○	○
L	N	(linked)		Timer Contacts		Room Stat	

Fully pumped system only
For sealed systems only
Heat exchanger material, copper
Integral frost protection
SEDBUK rating D

Remove links 3–4 and 5–6 as appropriate. Terminals 4–5 are linked internally. If connecting to a system incorporating 2 × 2-port motorized valves or a mid-position valve, terminal 1 is Live Out and terminal 2 switch Live In, after removing link.

Ariston Micro Genus combination

Wall mounted

Connect a voltage free room thermostat to the 2-way terminal block on the back of the control panel after removing link.

Fully pumped system only
For sealed systems only
Heat exchanger material, copper
Integral frost protection
SEDBUK rating D

Atmos Coopra compact N30 condensing

N30C – Boiler
N30B – System
N30K – Combination

Wall mounted

6	5	4	3	2	1
○	○	○	○	○	○
				Room Stat	

Fully pumped system only
Suitable for sealed systems only
Heat exchanger material, stainless steel
Integral frost protection
SEDBUK rating A

External sensor and template sensor for cylinder may be fitted.

Atmos Multi 24/80, 32/80 plus condensing combination

Wall mounted

Mains lead supplied. Connect 24 V room stat to terminals ZW on terminal block K1.

Fully pumped system only
Suitable for sealed systems only
Heat exchanger material, aluminum
Integral frost protection
SEDBUK rating A/B

Barlo Balmoral

See Halstead Balmoral

Barlo Blenheim 15/30, 30/40, 40/50, 50/60, 60/75

See Halstead Blenheim

Barlo Blenheim 42, 53

See Halstead 40H, 50H

Barlo Duo combination

Wall mounted

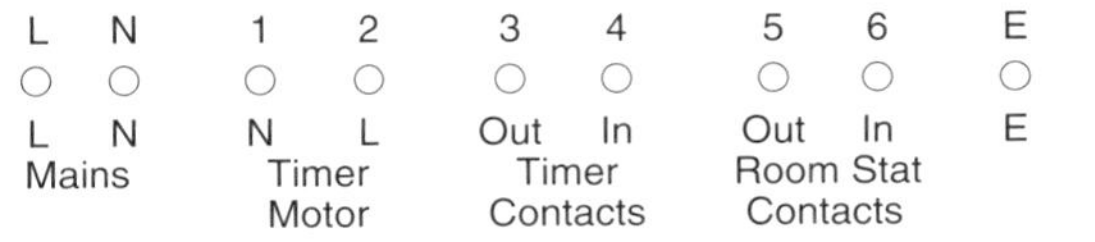

L	N	1	2	3	4	5	6	E
○	○	○	○	○	○	○	○	○
L	N	N	L	Out	In	Out	In	E
Mains		Timer Motor		Timer Contacts		Room Stat Contacts		

Suitable for sealed systems
Heat exchanger material, copper

If no timer or room thermostat link 3–4 and/or 5–6 as appropriate.

Baxi 100 HE condensing

Wall mounted

SL	E	N	PF
○	○	○	○
Switch Live	E	N	Pump Live (optional)

Fully pumped system only
Suitable for sealed systems
Heat exchanger material, aluminum
SEDBUK rating A

Baxi 401, 552

Back boiler unit

L	N	2
○	○	○
Switch Live	N	Perm* Live

Suitable for use on sealed systems using optional kit
Heat exchanger material, cast iron

*Permanent live only required if fire front has light bulbs, e.g. B, GF Super, LFE3 Super, VP and SP fire fronts.

Baxi Bahama 100 combination

Wall mounted

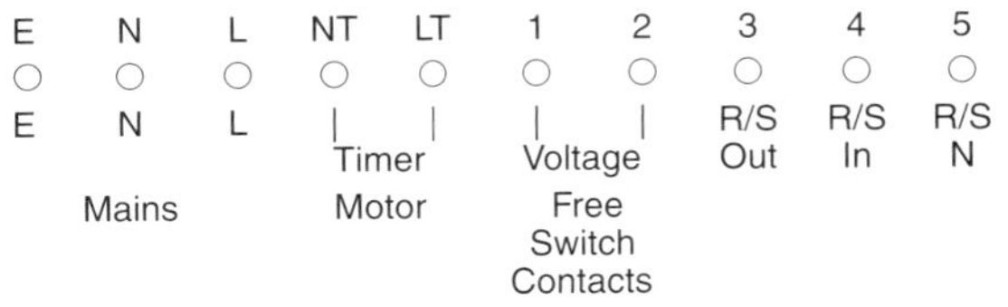

E	N	L	NT	LT	1	2	3	4	5
○	○	○	○	○	○	○	○	○	○
E	N	L	|	|	|	|	R/S Out	R/S In	R/S N
Mains			Timer Motor		Voltage Free Switch Contacts				

Fully pumped system only
For sealed systems only
Heat exchanger material, copper
Integral frost protection
SEDBUK rating D

Connect frost stat between 1 and 4.
Connect external time switch voltage free terminals across 1–2.

Baxi Barcelona condensing

Wall mounted

S/L	E	N	P/F
Switch Live	E	N	Pump Live

Fully pumped systems only
Suitable for sealed systems
Heat exchanger material, aluminium
SEDBUK rating A

The pump can be wired directly to the system or to terminal P/F. The P/F connection should only be used on a full TRV System without a by-pass.

Baxi Barcelona condensing system

Wall mounted

S/L	E	N	P/F
Switch Live	E	N	Pump Live

Fully pumped systems only
For sealed systems only
Heat exchanger material, aluminium
SEDBUK rating A

Terminal P/F is an optional pump feed which only needs to be used when fitting an additional external pump on a full TRV system.

Baxi Boston

Floor standing

With/without integral programmer

N	L	1	2	3	4	L1	N1
N Mains	L Mains	HW Off	CH Off	HW On	CH On	Switch Live	N

Not suitable for use on sealed systems
Heat exchanger material, cast iron

Programmer if fitted
Landis & Gyr RWB2

Baxi Boston 2, OF, RS

Floor standing

E	N	L
E	N	Switch Live

Not suitable for sealed systems
Heat exchanger material, cast iron

The pump can be wired directly to the system or to terminal P/F. The P/F connection should only be used on a full TRV System without a by-pass.

Baxi Combi 80e, 105e combination

Wall mounted

br	bl	g/y	bk	bk
	Mains		1 Out	2 In

(Terminals 1 and 2 linked.)

Fully pumped systems only
For sealed systems only
Heat exchanger material, copper
Integral frost protection
SEDBUK rating D

External controls such as time clock with voltage free contacts, or room thermostat suitable for use at 240V, should be connected to terminal block after removing link. Connect room stat anticipator, if fitted, to N.

Baxi Combi 80 Eco combination

Wall mounted

Fully pumped systems only
For sealed systems only
Heat exchanger material, copper
SEDBUK rating D

External controls such as time clock with voltage free contacts, or room thermostat suitable for use at 240V, should be connected to terminal block after removing link. Connect room stat anticipator, if fitted, to N.

Baxi Combi 80 Maxflue combination

Wall mounted

Fully pumped systems only
For sealed systems only
Heat exchanger material, copper
Integral frost protection
SEDBUK rating D

External controls such as time clock with voltage free contacts, or room thermostat suitable for use at 240V, should be connected to terminal block after removing link. Connect room stat anticipator, if fitted, to N.

Baxi Combi 130 HE combination

Wall mounted

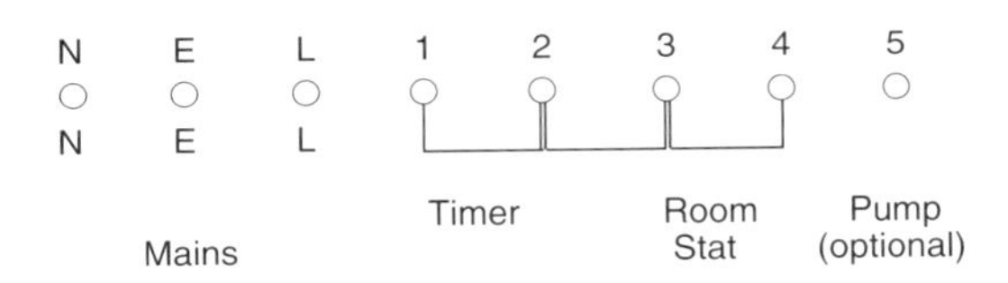

Fully pumped system only
For sealed systems only
Heat exchanger material, aluminium
Integral frost protection
SEDBUK rating A

Connect external time switch voltage free terminals across 1–2 after removing link.
Connect room stat across terminals 3–4 after removing link.
Terminal 5 can be used as a live feed for an external pump.

Baxi FS range

Floor standing

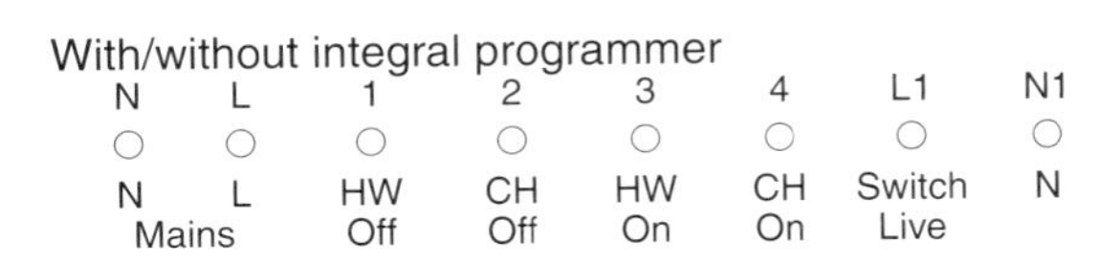

Not suitable for use on sealed systems
Heat exchanger material, cast iron

Programmer if fitted
Landis & Gyr RWB2

Baxi Genesis combination

Wall mounted

Fully pumped systems only
Suitable for sealed systems
Heat exchanger material, copper

External controls, such as time clock with voltage free contacts, or room thermostat suitable for use at 24V, should be connected between terminals 5 and 7 after removing link.

Baxi Maxflow Combi FS combination

Floor standing

br bl g/y

Mains

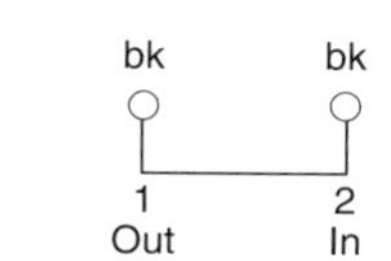

Fully pumped systems only
For sealed systems only
Heat exchanger material, copper
Integral frost protection
SEDBUK rating D

External controls such as time clock with voltage free contacts, or room thermostat suitable for use at 240V should be connected to terminal block after removing link. Connect room stat anticipator, if fitted, to N.

Baxi Maxflow Combi WM combination

Wall mounted

Fully pumped systems only
For sealed systems only
Heat exchanger material, copper
Integral frost protection
SEDBUK rating D

External controls such as time clock with voltage free contacts, or room thermostat suitable for use at 240V should be connected to terminal block after removing link. Connect room stat anticipator, if fitted, to N.

Baxi Solo PF

Wall mounted

PL PN PE — Pump
S/L — Switch Live
L N E — Mains

Fully pumped systems only
Suitable for use on sealed systems
Heat exchanger material, cast iron

Baxi Solo RS

Wall mounted

L — Switch Live
N — N
E — E

Not suitable for use on sealed systems
(see Solo RS/SS)
Heat exchanger material, cast iron

Baxi Solo RS/SS

Wall mounted

Fully pumped systems only
Suitable for use on sealed systems
Heat exchanger material, cast iron

Baxi Solo 2 PF

Wall mounted

Pump			E	N	PL	SWL
○	○	○	○	○	○	○
L	N	E	E	N	Perm Live	Switch Live

Fully pumped systems only
Suitable for use on sealed systems
Heat exchanger material, cast iron
Integral frost protection

Baxi Solo 2 RS

Wall mounted

Fully pumped systems

a)	SL	N	E	PL	L	N	E
	○	○	○	○	○	○	○
	Switch Live	N	E	Perm Live	Pump Live	N	E

Suitable for use on sealed systems
Heat exchanger material, cast iron

Gravity hot water, pumped central heating systems

b)	SL	N	E	PL	L	N	E
	○	○	○	○	○	○	○
	Switch Live	N	E		Not Used		

Suitable for use on sealed systems
Heat exchanger material, cast iron

When used on system b) the overheat thermostat should be by-passed by referring to installation instructions.

Baxi Solo 3 PFL

Wall mounted

Pump			E	N	PL	SW
○	○	○	○	○	○	○
L	N	E	E	N	Perm Live	Switch Live

Fully pumped system only
Suitable for sealed systems
Heat exchanger material, cast iron
Integral frost protection
SEDBUK rating D/E

Biasi Parva M90 combination

Wall mounted

Fully pumped system only
Suitable for sealed systems only
Heat exchanger material, copper/steel
Integral frost protection
SEDBUK rating D

Connect room stat between 3 and 1 after removing link.
Connect frost stat between 2 and 1.

Biasi Riva 24S, 28S, 24SR combination

Wall mounted

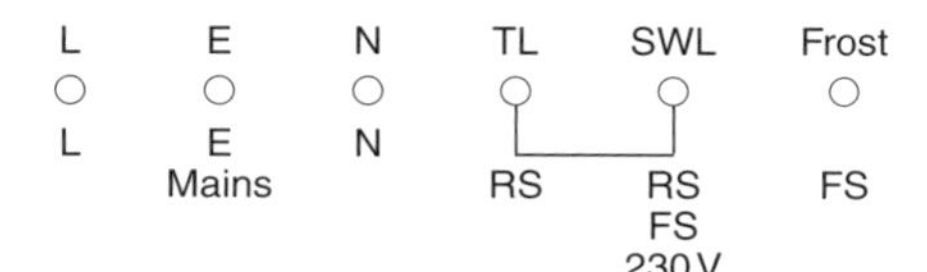

Fully pumped system only
Suitable for sealed systems only
Heat exchanger material, copper/steel
SEDBUK rating D

Remove link if fitting room stat. Room stat anticipator if fitted should be connected to N. Connect frost stat between SWL and Frost.

Bosch Greenstar condensing

Wall mounted

Fully pumped system only
Suitable for sealed systems only
Heat exchanger material, aluminium
SEDBUK rating A

Bosch RX2 condensing combination

Wall mounted

ST8
NS LS LR Spare
N L Switch Live

To external controls – 230V

Fully pumped system only
Suitable for sealed systems only
Heat exchanger material, copper
Integral frost protection
SEDBUK rating C

Boulter Buderus 600R condensing

Wall mounted

A room thermostat of either mains voltage or with voltage free connections can be wired to the boiler but not to the same terminals. A mains voltage room thermostat should be connected to terminals 1–2 of the 230V connection box. A voltage free room thermostat should be connected to terminals 1–2 of the volt free connection box. Boulter Buderus supply a range of temperature controls for the latter.

Fully pumped systems only
Suitable for sealed systems
Heat exchanger material, aluminium
SEDBUK rating A

Boulter Buderus 800 condensing

Wall mounted

When these boilers are installed with external controls, e.g. zone valves, mid-position valves, programmers and time clocks, the switched live should be connected to L on the relay pcb. No mains wiring should be connected to terminals 1–12.

Fully pumped systems only
Suitable for sealed systems
Heat exchanger material, aluminium
SEDBUK rating A

Broag-Remeha Selecta combination condensing

Wall mounted

A 24V room thermostat should be connected to terminals 2–3 on X4.
A range of manufacturers control equipment is available.

Fully pumped systems only
Suitable for sealed systems only
Heat exchanger material, aluminium
Integral frost protection
SEDBUK rating A

Broag-Remeha Selecta system condensing

Wall mounted

A 24V room thermostat should be connected to terminals 2–3 on X4.
A range of manufacturers control equipment is available.

Fully pumped systems only
Suitable for sealed systems only
Heat exchanger material, aluminium
Integral frost protection
SEDBUK rating A

Broag-Remeha Quinta condensing

Wall mounted

A range of manufacturers control equipment is available.
All connections are made to 24-way terminal strip X11, Live to terminal 16 and Neutral to terminal 18.

Fully pumped systems only
Suitable for sealed systems only
Heat exchanger material, aluminium
Integral frost protection
SEDBUK rating A

Chaffoteaux Britony combination

Wall mounted

Multi-pin Plug

Suitable for sealed systems
Heat exchanger material, copper
SEDBUK rating E

Connect room thermostat suitable for 24V into multi-pin plug removing link.

Chaffoteaux Britony SE combination

Wall mounted

A mains lead is supplied.
A24V room thermostat should be connected to multi-pin plug on terminal board after removing link.

Fully pumped systems only
Suitable for sealed systems only
Heat exchanger material, copper
Integral frost protection
SEDBUK rating D

Chaffoteaux Britony system II

Wall mounted

The System II Plus has separate water storage unit.

Fully pumped systems only
Suitable for sealed systems only
Heat exchanger material, copper
SEDBUK rating D

Chaffoteaux Calydra Comfort combination

Wall mounted

A mains lead is supplied.
A 24 V room thermostat should be connected to multi-pin plug on terminal board after removing link.

Fully pumped systems only
Suitable for sealed systems only
Heat exchanger material, copper
Integral frost protection
SEDBUK rating D

Chaffoteaux Calydra Green condensing combination

Wall mounted

Fully pumped systems only
Suitable for sealed systems only
Heat exchanger material, aluminium
Integral frost protection
SEDBUK rating D

Chaffoteaux Calydra Green system condensing

Wall mounted

Fully pumped systems only
Suitable for sealed systems only
Heat exchanger material, aluminium
Integral frost protection
SEDBUK rating A

Chaffoteaux Celtic combination

Wall mounted

L	N	4	5
○	○	○	○
L	N	Timer or Room Stat Contacts	

If no timer or room thermostat fit link 4–5

Suitable for sealed systems
Heat exchanger material, copper
Integral frost protection

Chaffoteaux Celtic Plus FF combination

L N 4 5

L N Timer or Room Stat Contacts

If no timer or room thermostat fit link 4–5

Wall mounted

Suitable for sealed systems
Heat exchanger material, copper

Chaffoteaux Centora Green combination condensing

Connect mains to multi-pin plug J1.
A24V room thermostat and/or timer with voltage free terminals can be connected to the multi-pin plug after removing link.

Wall mounted

Fully pumped systems only
Suitable for sealed systems only
Heat exchanger material, stainless steel
Integral frost protection
SEDBUK rating A

Chaffoteaux Challenger

L N N L 5 6

Mains Pump Live to Controls if req. Switch Live

Wall mounted

Fully pumped systems only
Suitable for use on sealed systems
Heat exchanger material, copper

Chaffoteaux Flexiflame 140

If no timer or room thermostat fit link 6–7

Wall mounted

Fully pumped systems only
Suitable for use on sealed systems
Heat exchanger material, copper
The integral pump is not the system pump

Chaffoteaux Minima MX2 combination

A mains lead is supplied.
A 24V room thermostat can be connected to the control board after removing link.

Wall mounted

Fully pumped systems only
Suitable for sealed systems only
Heat exchanger material, copper
SEDBUK rating D

Chaffoteaux Sterling OF and FF combination

Wall mounted

L1 N2 3 4 5
Mains

Suitable for use on sealed systems
Heat exchanger material, copper

External controls such as time clock with voltage free contacts, or room thermostat, should be connected between terminals 4 and 5. Terminal 3 can be used to connect room thermostat neutral.

Chaffoteaux Sterling SB system boiler

Wall mounted

Available with or without 3-port mid-position valve

Without valve – heating only:

Room Stat: N COM DEM
Mains: N L E

Suitable for use on sealed systems
Heat exchanger material, copper

Link N–N, COM–L only

With 3-port mid-position valve:
Connect valve as per colours indicated on terminal strip. Room thermostat and mains as above.

Mains: N L
Programmer: 5 4 3 2 1 (4: CH ON; 1: HW ON)
Cylinder Stat: 7 0 1 (SAT COM DEM)

Suitable for use on sealed systems
Heat exchanger material, copper

Link L–L, N–N, N–blue. Connections: White–room stat demand, orange–Cyl stat satisfied Prog 4-room stat common, prog 1-cyl stat common.

Combi Company GEM combination

Wall mounted

Mains: L N E
12V DC: L2 L1

Suitable for use on sealed systems
Heat exchanger material, copper

External controls such as a time clock with voltage free contacts or room thermostat should be connected between terminals L2 and L1. A frost thermostat should be wired across L2 and red wire connected to DHW flow switch.

Eco-hometec EC Compact condensing

A main lead is supplied.
A range of manufacturers control equipment is available.

Wall mounted

Fully pumped systems only
Suitable for sealed systems
Heat exchanger material – stainless steel
Integral frost protection
SEDBUK Rating A

Eco-hometec EC 16, 23, 31, 38 Condensing

Mains connection via 3-pin plug.
A range of manufacturers control equipment is available.

Wall mounted

Fully pumped systems only
Suitable for sealed systems
Heat exchanger material, stainless steel
Integral frost protection
SEDBUK rating A

Eco-hometec Solar Combi condensing combination

Mains connection via 3-pin plug.

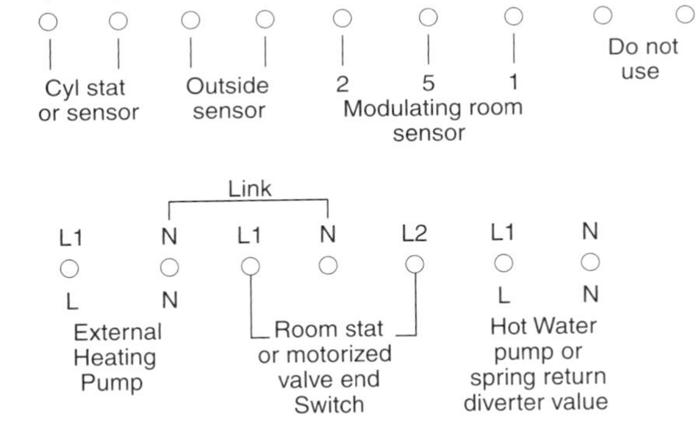

Wall mounted

Fully pumped systems only
Suitable for sealed systems
Heat exchanger material, stainless steel
Integral frost protection
SEDBUK rating A

ELM Leblanc GLM 5.20, 5.32 combination

Wall mounted

Suitable for sealed systems
Heat exchanger material, copper
The timer should have voltage free terminals.
The room thermostat should be suitable for low voltage (32V).

ELM Leblanc GVM 4.20 combination

Wall mounted

Suitable for sealed systems
Heat exchanger material, copper
The timer should have voltage free contacts.
Room thermostat should be suitable for mains voltage 240V.

ELM Leblanc GVM 7.23 combination GVM 7.28

Wall mounted

Timer
Room Stat
Mains 240V

Suitable for sealed systems
Heat exchanger material, copper
Timer and or room thermostat should be connected into their appropriate terminals and be suitable for 24V.

ELM Leblanc GVM 14/20 combination

Wall mounted

N L
240V Mains
Timer
Room Stat

Suitable for sealed systems
Heat exchanger material, copper

ELM Leblanc GVM C.21, C.23 condensing combination

Wall mounted

N L
240V Mains
Timer
Room Stat

Suitable for sealed systems
Heat exchanger material, copper

Timer should have voltage free contacts. Room thermostat should be suitable for low voltage 32V.

Eurocombi Styx combination

Wall mounted

Boiler is supplied with a 3-core mains lead
External controls such as time clock with voltage free contacts or room thermostat should be connected into connector C after removing either brown link.

Suitable or sealed systems
Heat exchanger material, copper

Fagor Eco Compact combination

Wall mounted

Mains lead supplied. External controls such as room stat on time switch should have voltage free terminals and be suitable for 24V. Connect to block CM2.

Fully pumped system only
Suitable for sealed systems only
Heat exchanger material, copper/stainless steel
Integral frost protection

Fagor FE-20E combination

Wall mounted

Mains lead supplied. A 24V room thermostat and/or time clock with voltage free contacts should be connected to 2-way connector block.

Fully pumped system only
Suitable for sealed systems only
Heat exchanger material
SEDBUK rating D

Ferroli 76 FF combination

24V L N E

Timer or Room stat — Mains

Wall mounted

Suitable for sealed systems
Heat exchanger material, copper
Room thermostat should be suitable for 24V.

If no timer or room thermostat fit link. Do not wire room thermostat neutral.

Ferroli 77 CF, FF and FF Popular combination

Wall mounted

Suitable for sealed systems
Heat exchanger material, copper
77 FF Popular does not include integral clock.

Ferroli 100 FF, 120 CF combination

L E N — 4 5

Mains — 24V

Wall mounted

Suitable for sealed systems
Heat exchanger material, copper

External controls such as time clock with voltage free contacts, or room thermostat suitable for use at 24V, should be connected between terminals 4 and 5 after removing link.

Ferroli Arena 30A system condensing

A voltage free room thermostat can be connected to terminals 1–2 on the 12-way connection block. Other controls including Eco/Comfort selection switch (3–4), external temperature sensor (7–8), remote control (9–10) and DHW storage sensor (11–12) can also be connected to this connection block.

Wall mounted

Fully pumped systems only
Suitable for sealed systems only
Heat exchanger material, aluminium
SEDBUK rating A

Ferroli Arena 30c combination condensing

A voltage free room thermostat can be connected to terminals 1–2 on the 12-way connection block. Other controls including remote control (9–10), external temperature sensor (7–8), Eco/Comfort selection switch (3–4) and domestic hot water exclusion switch (5–6) can also be connected to this connection block.

Wall mounted

Fully pumped systems only
Suitable for sealed systems only
Heat exchanger material, aluminium
SEDBUK rating A

Ferroli Domina 80E combination

Wall mounted

A mains supply should be connected to the 3-way terminal board XI. Voltage free external controls and/or a 24V room thermostat should be connected to the 2-way terminal board (terminals 3 & 4).

Fully pumped systems only
Suitable for sealed systems only
Heat exchanger material, copper
SEDBUK rating D

Ferroli F24, F30 combination

Wall mounted

A mains supply should be connected to the 3-way terminal board XI. Voltage free external controls and/or a 24V room thermostat should be connected to the 2-way terminal board (terminals 3 & 4).

Fully pumped systems only
Suitable for sealed systems
Heat exchanger material, copper
SEDBUK rating D

Ferroli Logica condensing combination

Wall mounted

L E N 4 5
Mains 24V

Suitable for sealed systems
Heat exchanger material, copper

External controls such as time clock with voltage free contacts, or room thermostat suitable for use at 24V, should be connected between terminals 4 and 5 after removing link.

Ferroli Modena 80E combination

Wall mounted

A mains supply should be connected to the 3-way terminal board XI. Voltage free external controls and/or a 24V room thermostat should be connected to the 2-way terminal board (terminals 3 & 4).

Fully pumped systems only
Suitable for sealed systems only
Heat exchanger material, copper
SEDBUK rating D

Ferroli Nouvelle Elite 127/92 combination

Wall mounted

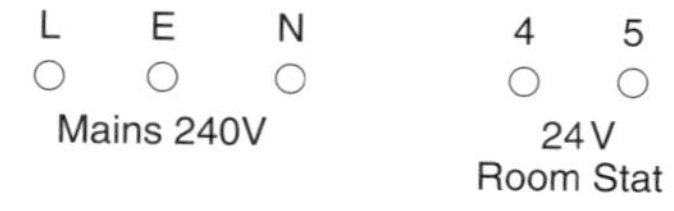

Suitable for sealed systems
Heat exchanger material, copper

External controls such as time clock, or room thermostat suitable for use at 24V, should be connected between terminals 4 and 5 after removing link. Do not wire room thermostat neutral.

Ferroli Optima combination

Wall mounted

L	E	N	4	5
	Mains		24V	

Suitable for sealed systems
Heat exchanger material, copper

External controls such as time clock with voltage free contacts, or room thermostat suitable for use at 24V, should be connected between terminals 4 and 5 after removing link.

Ferroli Optima 201–1001

Wall mounted

Mains supply to X7, 1–2. A 24V thermostat can be connected to the 2-way terminal board (4 & 5) after removing link.

Fully pumped systems only
Suitable for sealed systems only
Heat exchanger material, copper
Integral frost protection
SEDBUK rating E

Ferroli Optima 2001 condensing

Wall mounted

Fully pumped systems only
Suitable for sealed systems only
Heat exchanger material, aluminium & copper
Integral frost protection
SEDBUK rating B

Mains supply to X7, 1–2. A 24V thermostat can be connected to the 4-way terminal board as below after removing link.

Ferroli Roma

Wall mounted

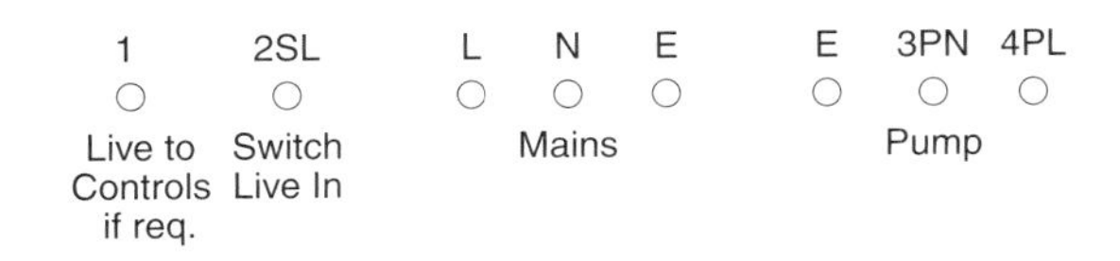

Fully pumped systems only
Suitable for use on sealed systems
Heat exchanger material, copper
Integral frost protection

Ferroli Sigma

Wall mounted

SWL	L	E	N	PL	E	PN
Switch Live	Perm Live	E	N	Pump Live	E	Pump N

Suitable for sealed systems
Heat exchanger material, cast iron
SEDBUK rating D/E

For pumped CH and gravity HW, move link on X8 from CP (fully pumped) to GC.

Ferroli Sys 10–23 system

Wall mounted

7	6	5	4	3	2	1
○	○	○	○	○	○	○
Switch Live				E	N	L

Room thermostat should be suitable for mains voltage

Fully pumped systems only
Suitable for sealed systems only
Heat exchanger material, copper
SEDBUK rating D

Ferroli Tempra 24 combination

Wall mounted

A mains supply should be connected to the 3-way terminal board XI. Voltage free external controls and/or a 24V room thermostat should be connected to the 2-way terminal board (terminals 3 & 4).

Fully pumped systems only
Suitable for sealed systems only
Heat exchanger material, copper
SEDBUK rating D/E

Ferroli Xignal combination

Wall mounted

1	2	3	4	5	6	7	8	9	10
○	○	○	○	○	○	○	○	○	○
L	E	N		Room Sensor		External Sensor			
	Mains								

Suitable for sealed systems
Heat exchanger material, copper
Both sensors are supplied with the boiler.
Computerized boiler with synthesized voice communication.

Gemini 960E combination

Wall mounted

B4	E	T2	T1	N	E	L
○	○	○	○	○	○	○
		240V Room Stat		240V Mains		

Suitable for sealed systems
Heat exchanger material, copper

Electrical connection via 7-pin plug as shown. Connect frost thermostat across B4-T1 in plug.

Geminox MZ condensing

Wall mounted

P	P	TH	TH	N	N	PH
○	○	○	○	○	○	○
Link		Timer or Room stat contacts		N	Mains	

Link P–P to be removed if fitting digi control

Fully pumped systems only
Suitable for sealed systems
Heat exchanger material, aluminium

Glotec GT80 condensing

Wall mounted

10	9	8	7	6	5	4	3	2	1
○	○	○	○	○	○	○	○	○	○
Switch Live	Live to contacts if req.	N	E	E	L	N	N	L	E
				Pump (optional connection)			Mains		

Fully pumped systems only
Suitable for use on sealed systems
Heat exchanger material, stainless steel

Glow-Worm 18 si, 30 si system

Wall mounted

Connection of external controls will be different depending on whether they are voltage free or mains voltage.
Voltage free – remove wire link from 2-way connection on control box cover.
Mains voltage – refer to diagram below which shows the controls interface PCB.

○ – 230V switched live from optional frost stat
○ – 230V switched live from heating controls
○ – Do not connect

Fully pumped system only
Suitable for sealed systems only
Heat exchanger material, stainless steel
Integral frost protection
SEDBUK rating D

If mains voltage external controls are used, the mains voltage heating controls plug should be installed on the controls interface PCB.

Glow-Worm 18 sxi, 30 sxi condensing system

Wall mounted

○ – 230V switched live from optional frost stat
○ – 230V switched live from heating controls
○ – Do not connect

Fully pumped system only
Suitable for sealed systems only
Heat exchanger material, stainless steel
Integral frost protection
SEDBUK rating A

Remove link if fitting voltage-free or mains external controls. If the link is not removed, the boiler will run continuously.

Glow-Worm 24 cxi, 30 cxi, 38 cxi condensing combination

Wall mounted

Connection of external controls will be different depending on whether they are voltage free or mains voltage.
Voltage free – remove wire link from two-way connection on control box cover.
Mains voltage – refer to diagram below which shows the controls interface PCB.

○ – 230V switched live from optional frost stat
○ – 230V switched live from heating controls
○ – Do not connect

Fully pumped system only
Suitable for sealed systems only
Heat exchanger material, stainless steel
Integral frost protection
SEDBUK rating A

If mains voltage external controls are used, the mains voltage heating controls plug should be installed on the controls interface PCB.

Glow-Worm 30 Ci, 35 Ci combination

Wall mounted

Connection of external controls will be different depending on whether they are voltage free or mains voltage.
Voltage free – remove wire link from-way connection on control box cover.
Mains voltage – refer to diagram below which shows the controls interface PCB.

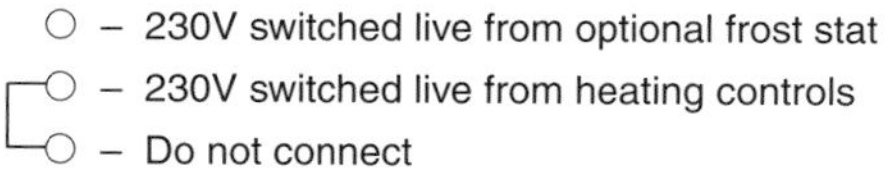

○ – 230V switched live from optional frost stat
○ – 230V switched live from heating controls
○ – Do not connect

Fully pumped system only
Suitable for sealed systems only
Heat exchanger material, stainless steel
Integral frost protection
SEDBUK rating D

If mains voltage external controls are used, the mains voltage heating controls plug should be installed on the controls interface PCB.

Glow-Worm 45 and 56B

Back boiler unit

N	L	N	SL
○	○	○	○
N	L*	N	Switch Live

Not suitable for use on sealed systems
Heat exchanger material, cast iron
SEDBUK rating D

*Permanent live only required if fire front has light bulbs, e.g. Homeglow and LFC.

Glow-Worm Compact 70e, 80e combination

Wall mounted

L	N	E	R1	R2
○	○	○	○	○
L	N	E	External Controls	
Mains				

Fully pumped system only
Suitable for sealed systems only
Heat exchanger material, copper
SEDBUK rating D

Connect external controls suitable for mains voltage to R1–R2 after removing link.

Glow-Worm Compact 60, 100 system

Wall mounted

LP	N	E	LS
○	○	○	○
Mains			Switch Live

Fully pumped system only
Suitable for sealed systems only
Heat exchanger material, copper
Integral frost protection
SEDBUK rating D

Glow-Worm Economy

Wall mounted

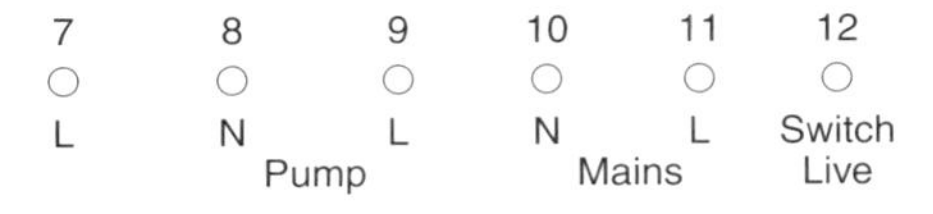

Remove link 7–12

Fully pumped systems only
Suitable for use on sealed systems
Heat exchanger material, copper

Glow-Worm Energysaver condensing

Wall mounted

N L 1 PN PL SL

Mains Pump Switch Live

Remove link 7–12

Fully pumped systems only
Suitable for use on sealed systems
Heat exchanger material, copper
SEDBUK rating B

Glow-Worm Energysaver 30e, 40e, 80e condensing

Wall mounted

Remove link 1–SL

Fully pumped system only
Suitable for sealed systems
Heat exchanger material, copper
SEDBUK rating B

Glow-Worm Energysaver Combi 2 80, 100 condensing combination

Wall mounted

Black control Plug

Red Link Blue Link

Fully pumped system only
Suitable for sealed systems only
Heat exchanger material, aluminum
Integral frost protection
SEDBUK rating A

Mains voltage must not be connected to the black control plug.
Connect room stat after removing red link.
Connect timer after removing blue link.
Connect frost stat across two outer terminals if either red or blue link removed.

Glow-Worm Express combination

Wall mounted

5-pin internal plug

Suitable for sealed systems
Heat exchanger material, copper

Glow-Worm Fuelsaver 'B' MK2

Wall mounted

7	8	9	10	11	12
L	N	L	N	L	Switch Live
Pump		Mains			

Remove link 7–12

Fully pumped systems only
Suitable for use on sealed systems
Heat exchanger material, copper

Glow-Worm Fuelsaver 'BR' MK2

Wall mounted

E	N	L	SL	9	8	7
	Mains		Switch Live	L	N	L
				Pump		

Remove link SL–9

Fully pumped systems only
Suitable for use on sealed systems
Heat exchanger material, copper

Glow-Worm Fuelsaver Complheat condensing

Wall mounted

Electrical connection of switch live (L), neutral (N), and earth (E), via 3-pin internal plug

Fully pumped systems only
Suitable for ealed systems
Heat exchanger material, copper
SEDBUK rating E

Glow-Worm Fuelsaver 'F'

Wall mounted

7	8	9	10	11	12
L	N	L	N	L	Switch Live
Pump		Mains			

Remove link 7–12

Fully pumped systems only
Suitable for use on sealed systems except 100F
Heat exchanger material, copper

Glow-Worm Fuelsaver 'R' MK2

Wall mounted

E	N	L	SL	9	8	7
E	N	L	Switch Live	L	N	L
	Mains			Pump		

Remove link SL–9

Fully pumped systems only
Suitable for use on sealed systems
except 75R MK2
Heat exchanger material, copper

Glow-Worm Fuelsaver UFB

Wall mounted

E	N	L	SL	9	8	7
E	N	L	Switch Live	L	N	L
	Mains			Pump		

Remove link SL–9

Fully pumped systems only
Suitable for use on sealed systems except 100F
Heat exchanger material, copper

Glow-Worm Hideaway

Floor standing

Electrical connection via 3-pin internal plug

Not suitable for use on sealed systems
Heat exchanger material, cast iron
SEDBUK rating D/E

Glow-Worm Micron 30FF

Wall mounted

LS ○ Switch Live
N ○ N
E ○ E

Fully pumped system only
Suitable for sealed systems only
Heat exchanger material, copper
Integral frost protection
SEDBUK rating D

Glow-Worm Micron 60FF

Wall mounted

LP ○ Perm Live
LS ○ Switch Live
N ○ N
E ○ E
E ○ E
PL ○ L (Pump)
PN ○ N (Pump)

Fully pumped system only
Suitable for sealed systems
Heat exchanger material, copper
Integral frost protection
SEDBUK rating D

Glow-Worm Micron 100FF, 120FF

Wall mounted

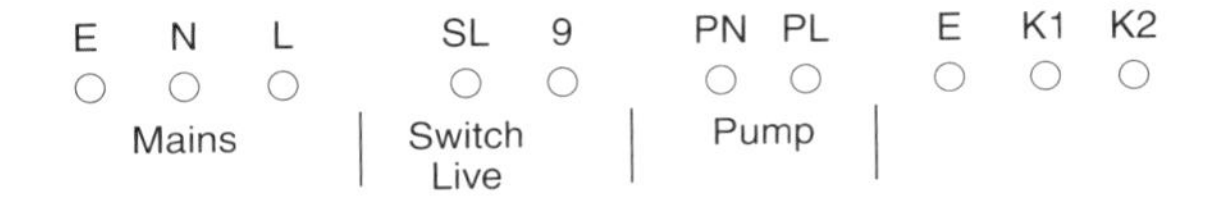

Fully pumped system only
Suitable for sealed systems
Heat exchanger material, stainless steel
SEDBUK rating D

Remove link SL–9 when fitting external controls. Fit link K1–K2 on open vented systems only.

Glow-Worm Spacesaver 75 CF

Wall mounted

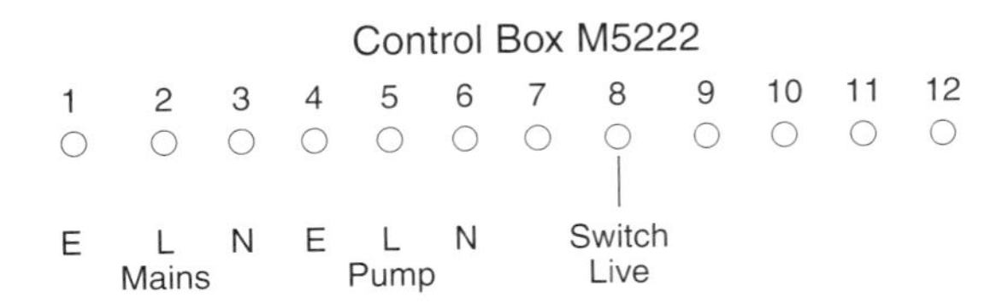

Remove link 7–8

Fully pumped systems only
Not suitable for use on sealed systems
Heat Exchange materials, cast iron

Glow-Worm Spacesaver 'BR' MK2

Wall mounted

E ○ E
N ○ N
L ○ Switch Live

80BR MK2 – Fully pumped systems only
Not suitable for use on sealed systems
Heat exchanger material, cast iron

Glow-Worm Spacesaver 'F'

Wall mounted

L	N	E
○	○	○
Switch Live	N	E

Not suitable for use on sealed systems
Heat exchanger material, cast iron

Glow-Worm Spacesaver 'KFB'

Wall mounted

5	6	7	8	9	10	11	12
○	○	○	○	○	○	○	○
E	E	L	N	L	N	L	Switch Live
			Pump		Mains		

Remove link 7–12

Fully pumped systems only
Not suitable for use on sealed systems
Heat exchanger material, cast iron

Glow-Worm Spacesaver 'R' MK2

Wall mounted

E	N	L
○	○	○
E	N	Switch Live

Not suitable for use on sealed systems
Heat exchanger material, cast iron

Glow-Worm Spacesaver 'RF'

Wall mounted

L	N	E
○	○	○
Switch Live	N	E

Not suitable for use on sealed systems
Heat exchanger material, cast iron

Glow-Worm Swiftflow combination

Wall mounted

Electrical connections are via a 5-pin plug

L	N	E	2	1
○	○	○	○	○
L	N	E	L	L
	Mains		Out	In

Suitable for sealed systems
Heat exchanger material, copper
SEDBUK rating E/F

External controls can be connected between terminal 2 and 1 or directly into terminal 1 after removing link.

Glow-Worm Ultimate BF and CF

Wall mounted

L	N	E
○	○	○
Switch Live	N	E

80 BF – Fully pumped systems only
Not suitable for use on sealed systems
Heat exchanger material, cast iron
SEDBUK rating E/F

Glow-Worm Ultimate BFSS

Wall mounted

P	PN	E	4	SL	L	N	E	9	10
○	○	○	○	○	○	○	○	○	○
	Pump			Switch Live	Perm Live	N	E	Internal	

Remove link 4–SL if wiring external controls

Fully pumped systems only
Suitable for use on sealed systems
Heat exchanger material, cast iron
SUDBUK rating E/F

Glow-Worm Ultimate FF (excl. 80FF – see below)

Wall mounted

Gravity hot water, pumped central heating

E	N	L	SL	9	PN	PL	E	K1	K2
○	○	○	○	○	○	○	○	○	○
E	N		Switch Live					Link	

Fully pumped systems

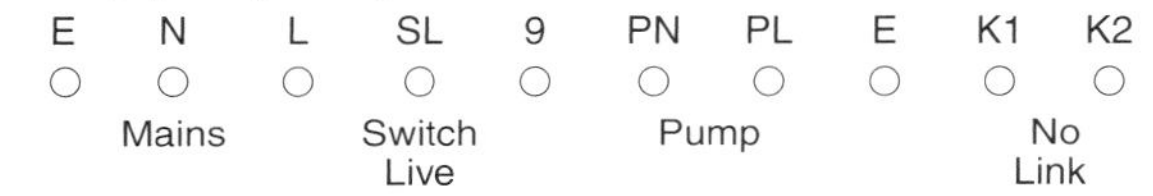

Suitable for use on sealed systems
Heat exchanger material, cast iron
SEDBUK rating D/E

Glow-Worm Ultimate 80FF

Wall mounted

E	7	8	9	SL	L	N	E
○	○	○	○	○	○	○	○
E	L Pump	N		Switch Live	Perm Live	N	E

Remove link 9–SL when wiring external controls

Fully pumped systems only
Not suitable for use on sealed systems
Heat exchanger material, cast iron
SEDBUK rating E

Glow-Worm Xtrafast 96, 120 combination

Wall mounted

Remove link to connect external voltage free control

Fully pumped system only
Suitable for sealed systems only
Heat exchanger material, copper
Integral frost protection
SEDBUK rating D

Glow-Worm Xtramax combination

Wall mounted

Remove link to connect external voltage free control

Fully pumped system only
Suitable for sealed systems only
Heat exchanger material, copper
Integral frost protection
SEDBUK rating D

Halstead 40H–50H

Wall mounted

E	L	N	NP	LP	EP	1	2	3	4
E	L	N		Pump		L	Switch Live	Internal Wiring	
	Mains								

Fully pumped systems only
Not suitable for use on sealed systems
Heat exchanger material, copper

Halstead 45F, 65F

As Halstead Balmoral

Halstead ACE & ACE High combination

Wall mounted

L3	N	E	L2	L1
	N	E	Mains Live	

Fully pumped system only
Suitable for sealed systems only
Heat exchanger material, copper
Integral frost protection
SEDBUK rating D

Remove link between L1 and L3 if fitting external controls, which must be suitable for 240V and voltage free.

Halstead Balmoral

Wall mounted

12	11	10	9	8	7	6	5	4	3	2	1
L	N	E	L	N	L	Switch Live			Internal Wiring		
	Mains		Pump								

Terminals 7 and 12 are linked

Fully pumped systems only
Suitable for use on sealed systems
Heat exchanger material, copper

Halstead Bentley

Wall mounted

E	N	L
E	N	Switch Live

Suitable for use on sealed systems using optional kit
Heat exchanger material, cast iron

Halstead Best

See Halstead Boss – renamed

Halstead Blenheim

Wall mounted

4	3	2	1	EP	LP	NP	N	L	E
Internal Wiring		Switch Live	L	E	L	N	N	L	E
					Pump			Mains	

Fully pumped systems only
Suitable for use on sealed systems (15/30 using optional kit)
Heat exchanger material, copper

Halstead Boss

Wall mounted

L N E 1 2
Mains — Switch Live, Pump Live

Suitable for use on sealed systems
Heat exchanger material, cast iron
SEDBUK rating D

A permanent live and pump live must be wired into the boiler for all systems including gravity hot water. The plug at the front of the control box should be positioned as necessary.

Halstead Buckingham and Buckingham 2

Floor standing

E N L1
E N Switch Live

Not suitable for use on sealed systems
Heat exchanger material, cast iron

Halstead Buckingham 3

Floor standing

Connect switched live, neutral and earth to 3-way connector (LS, N, E)

Not suitable for sealed systems
Heat exchanger material – cast iron
SEDBUK rating D

Halstead Buckingham 4

Floor standing

Connect switched live, neutral and earth to 3-way connector (LS, N, E)

Not suitable for sealed systems
Heat exchanger material, cast iron
SEDBUK rating D

Halstead Eden Cb condensing combination

Wall mounted

Connection of external controls will be different depending on whether they are voltage free or mains voltage.
Voltage free – remove wire link from 2-way connection on control box cover
Mains voltage – refer to diagram below which shows the controls interface PCB

- 230V switched live from optional frost stat
- 230V switched live from heating controls
- Do not connect

Fully pumped system only
Suitable for sealed systems only
Heat Exchanger material, stainless steel
Integral frost protection
SEDBUK rating A

If mains voltage external controls are used, the mains voltage heating controls plug should be installed on the controls interface PCB.

Halstead Eden Sb condensing system

Wall mounted

A mains lead is supplied. When connecting external mains voltage controls, remove red link from connection plug and connect as below.

- 230V switched live from optional frost stat
- 230V switched live from heating controls
- Do not connect

Fully pumped system only
Suitable for sealed systems only
Heat exchanger material, stainless steel
Integral frost protection
SEDBUK rating A

Halstead Eden Vb condensing

Wall mounted

1	2	3	4	5	6	7	8
Mains Live	Switch Live	Frost stat Live in	N	E	E	N	Pump Live

Link 1–2 if no external controls

Fully pumped system only
Suitable for sealed systems
Heat exchanger material, stainless steel
Integral frost protection
SEDBUK rating A

Halstead Finest, & Finest Gold combination

Wall mounted

Fully pumped system only
Suitable for seale d systems only
Heat exchanger material, copper
Integral frost protection
SEDBUK rating D

Remove link if fitting external controls, which must be suitable for 240V and voltage free.

Halstead Hero

Wall mounted

L2	L3	E	N	L1
Pump Live	Switch Live	E	N	Mains Live

Fully pumped system only
Suitable for sealed systems
Heat exchanger material, cast Iron
Integral frost protection
SEDBUK rating D

Halstead Quattro combination

Wall mounted

Room thermostat should be suitable for 240V.

Suitable for sealed systems
Heat exchanger material, copper
Integral frost protection

Halstead Quattro Gold combination

Wall mounted

L N E 1 2
Mains Room Stat

Room thermostat should be suitable for 240V.

Suitable for sealed systems
Heat exchanger material, copper
Integral frost protection

Halstead Trio combination

Wall mounted

Connection via internal plug

L N E 41 40
Mains Room Stat 24V

If no room thermostat link 40–41. Time clock integral to boiler.

Suitable for sealed systems
Heat exchanger material, copper

Heatline Compact combination

Wall mounted

A mains lead is supplied

Fully pumped system only
Suitable for sealed systems only
Heat exchanger material, copper
Integral frost protection

Heatline Compact system

Wall mounted

A mains lead is supplied

Fully pumped system only
Suitable for sealed systems only
Heat exchanger material, copper
Integral frost protection

Heatline Solaris condensing

Wall mounted

Fully pumped system only
Suitable for sealed systems only
Heat exchanger material, copper
Integral frost protection

Ideal C80FF, C95FF combination

Wall mounted

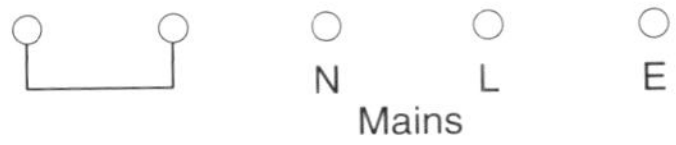

N L E
Mains

Fully pumped system only
Suitable for sealed systems only
Heat exchanger material, copper/stainless steel
SEDBUK rating D

Remove link in main terminal box when wiring external controls.
Remove link in secondary terminal box when wiring optional or external timer.

Ideal Classic BF

Wall mounted

N	E	L	
○	○	○	○
N	E	Switch Live	Int.

Suitable for use on sealed systems using optional kit
Heat exchanger material, cast iron
SEDBUK rating D/E

Ideal Classic FF

Wall mounted

Electrical connection via 3-pin plug

Not suitable for use on sealed systems using optional kit
Heat exchanger material, cast iron
SEDBUK rating D/E

Ideal Classic LX

As Classic with glass front panel and enhanced casing trim.

Ideal Classic system boiler

Wall mounted

Electrical connection via 3-pin plug

Fully pumped systems only
Suitable for sealed systems
Heat exchanger material, cast iron

Ideal Classic Combi NF80 combination

Wall mounted

CH	RS1	RS2	L1	CL	HW	E	L	N	CN
○	○	○	○	○	○	○	○	○	○
	Room Stat		L	L			Mains		N

Suitable for sealed systems
Heat exchanger material, cast iron

Fit link L1–HW. Connect room thermostat between terminals RS1 (out) and RS2 (in) after removing link. Connect external timer between terminals CL (live) and CH (control). Terminal CN is for room thermostat and timer neutral. A programmer can be installed as above with hot water channel connected to HW after removing link L1–HW.

Ideal Compact Extra system boiler

Wall mounted

L	N	E	PN	PL	DVN	DVL	E	1	2	A	B
○	○	○	○	○	○	○	○	○	○	○	○
	Mains supplied		Pump Integral			Diverter Valve Integral		Room Sensor		Cylinder Sensor	

Suitable for use on sealed systems using optional kit
Heat exchanger material, copper

Boiler incorporates a priority valve and a link to alter from HW to CH priority if required. Sensors could be wired in bell wire.

Ideal Elan 2CF, F, RS

LG	NG	LP	NP	L	N
○	○	○	○	○	○
Gas	Valve	Optional pump connection		Switch Live	N

Wall mounted

Fully pumped systems only
Suitable for use on sealed systems
Heat exchanger material, copper

Ideal Elan 2CF, NF, RS

○	○	○	○	○
Link for Optional Overheat Thermostat		Switch Live	E	N

Wall mounted

Fully pumped systems only
Suitable for use on sealed systems
Heat exchanger material, copper
See also Elan 2CF, F, RS

Ideal Icos condensing

L3	N	E	L2	L1
○	○	○	○	○
Perm Live	N	E	Switch Live	

Wall mounted

Fully pumped system only
Suitable for sealed systems
Heat exchanger material, aluminum
SEDBUK rating A

Connection via 5-way plug. A permanent live is essential. Remove link L1 and L2 if connecting external controls. A control circuit can be between L1 and L2 instead of a switched live into L2.

Ideal Icos system condensing

L3	N	E	L2	L1
○	○	○	○	○
L	N	E	Switch Live External	Live Out Controls
Mains				

Connect frost stat across L3–L2

Wall mounted

Fully pumped system only
Suitable for sealed systems
Heat exchanger material, aluminum
SEDBUK rating A

Ideal Imax condensing

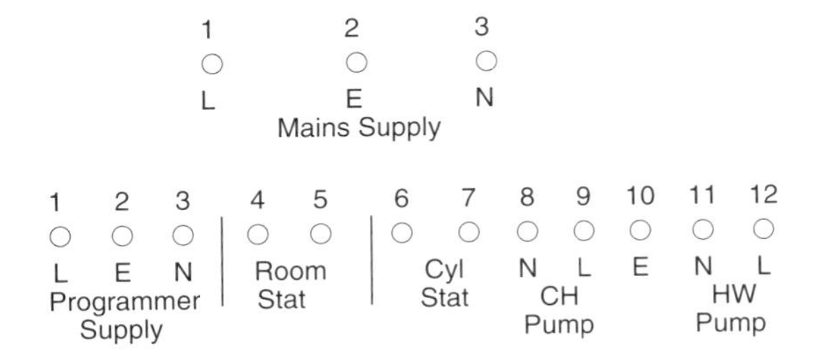

Wall mounted

Fully pumped systems only
Suitable for sealed systems
Heat exchanger material, aluminium
Integral frost protection
SEDBUK rating A

A permanent live is required. A diverter valve can be connected to 11–12 instead of a HW pump. Other controls including a cylinder sensor kit, outside temperature sensor and modulating sequencer kits can be incorporated into the controls.

Ideal Isar condensing combination

Wall mounted

L3	N	E	L2	L1
○	○	○	○	○
Perm Live	N	E	Switch Live	

Fully pumped systems only
Suitable for sealed systems only
Heat exchanger material, aluminium
SEDBUK rating A

Connection via 5-way plug. A permanent live is essential. Remove link L1–L2 if connecting external controls. A control circuit can be between L1 and L2 instead of a switched live into L2.

Ideal Mexico Slimline 2

Floor standing

Electrical connection via 3-pin internal plug

Not suitable for use on sealed systems
Heat exchanger material, cast iron
SEDBUK rating F

Ideal Mexico Super 2

Floor standing

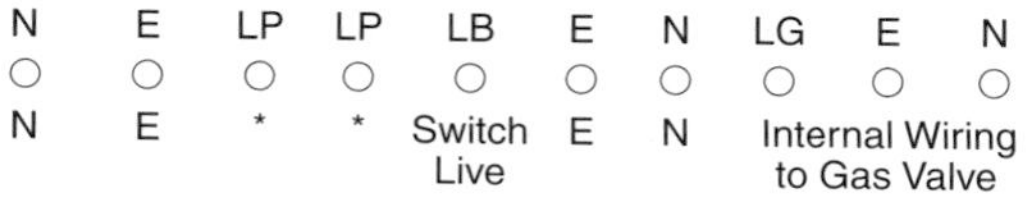

N	E	LP	LP	LB	E	N	LG	E	N
○	○	○	○	○	○	○	○	○	○
N	E	*	*	Switch Live	E	N	Internal Wiring to Gas Valve		

Not suitable for use on sealed systems
Heat exchanger material, cast iron
SEDBUK rating D/E

*Terminals LP are provided for connecting pump if convenient to do so.

Ideal Mexico Super FF

Floor standing

LH	LG	E	N	L	E
○	○	○	○	○	○
(linked)	(linked)	E	N	Switch Live	E

Suitable for sealed systems with kit
Heat exchanger material, cast iron
SEDBUK rating D/E

Ideal Minimiser condensing

Wall mounted

Requires a switched live and neutral connection.

Fully pumped system only
Suitable for sealed systems
Heat exchanger material, aluminum
SEDBUK rating B

Ideal Minimiser FF condensing

Wall mounted

A switched live and a neutral are connected into the 2-terminal connector block.

Fully pumped systems only
Suitable for use on sealed systems
Heat exchanger material, aluminium

Ideal Response 80, 100, 120 combination

Wall mounted

E N L

Mains

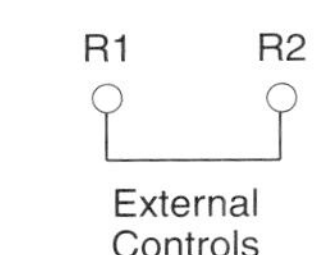

Fully pumped system only
For sealed systems only
Heat exchanger material, copper/cast iron
SEDBUK rating D

Connect external controls, suitable for 240V, across terminals R1–R2 after removing link.

Ideal Response FF combination

Wall mounted

Suitable for sealed systems
Heat exchanger material, copper and cast iron
SEDBUK rating D

External controls such as time clock with voltage free contacts, or room thermostat, should be connected between terminals R1 and R2 after removing link. Room thermostat neutral can be connected to N.

Ideal Response SE condensing combination

Wall mounted

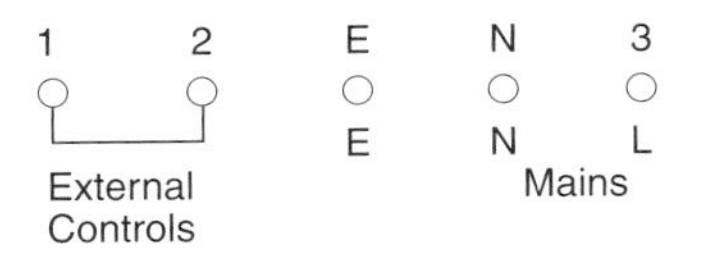

Fully pumped system only
For sealed systems only
Heat exchanger material, aluminium
Integral frost protection
SEDBUK rating B

Connect external controls across 1–2 after removing link.

Ideal Sprint 80F combination

Wall mounted

LP N E L1 N L2 L0 L E N

Pump N Mains

Suitable for sealed systems
Heat exchanger material, copper

Ideal Sprint RS75 combination

Wall mounted

Suitable for sealed systems
Heat exchanger material, copper

External controls such as time clock with voltage free contacts, or room thermostat, should be connected between terminals L1 and L2. Connect frost thermostat across L0–L2. Thermostats should be suitable for 240V.

Ideal Sprint Rapide RS 75N combination

Wall mounted

L N E N CK3 SCK4 RS1 RS2

240V Mains — Switch Live — Remove Link

Connect frost thermostat across FRS1–FRS2

Suitable for sealed systems
Heat exchanger material, copper

Ideal Sprint Rapide 90NF combination

Wall mounted

(a) Permanent Mains connection required
(b) Connect room thermostat between RS1-RS2
(c) Connect timer switch live to SCK4 (remove link CK3–SCK4)
(d) Connect frost thermostat between FRS1 and FRS2

Suitable for sealed systems
Heat exchanger material, copper

Ideal Systemiser SE condensing system

Wall mounted

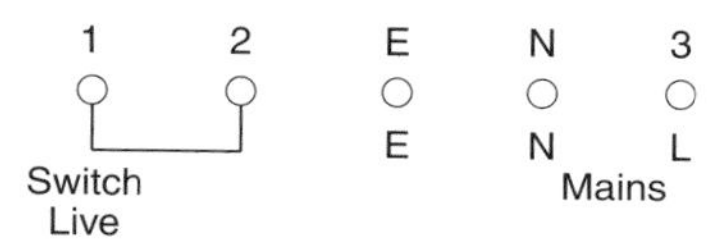

Connect room stat between 1–2 after removing link

Fully pumped system only
For sealed systems only
Heat exchanger material, aluminium
Integral frost protection
SEDBUK rating B

Ideal Turbo 2 condensing

Wall mounted

LP NP LB N L NT CH ON HW ON CH OFF HW OFF

Optional Pump — Switch Live — N L Mains* — N — Programmer Terminals If fitted

Fully pumped systems only
Suitable for use on sealed systems
Heat exchanger material, aluminium
SEDBUK rating B

*Permanent live is only required if internal programmer fitted.

Ideal W2000

Wall mounted

Electrical connection via 3-pin internal plug

Suitable for use on sealed systems using optional kit
Heat exchanger material, cast iron

Jaguar combination

Mains lead supplied. A voltage free room thermostat should be connected to the integral external controls connection after removing the link.

Wall mounted

Fully pumped system only
Suitable for sealed systems only
Heat exchanger material, copper/stainless steel
Integral frost protection
SEDBUK rating D

Johnson & Starley Reno HE 30C condensing combination

Wall mounted

Fully pumped system only
Suitable for sealed systems only
Heat exchanger material, stainless steel
Integral frost protection
SEDBUK rating A

Johnson & Starley Reno HE 25S condensing system

Wall mounted

Fully pumped system only
Suitable for sealed systems only
Heat exchanger material, stainless steel
Integral frost protection
SEDBUK rating A

Johnson & Starley Reno HE 25H condensing

Wall mounted

Fully pumped system only
Suitable for sealed systems only
Heat exchanger material, stainless steel
Integral frost protection
SEDBUK rating A

Keston 50, 60, 80 condensing

1	2	3	4	5	6	7	8	9	10	11	12
○	○	○	○	○	○	○	○	○	○	○	○
E	N	Switch Live									

Wall mounted

Fully pumped systems only
Suitable for use on sealed systems
Heat exchanger material, stainless steel
SEDBUK rating B

Keston Celsius 25 condensing

PL	N	E	SL	SL	LO	RUN
○	○	○	○	○	○	○
Perm Live	N	E	Switch Live			

Wall mounted

Fully pumped system only
Suitable for sealed systems
Heat exchanger material, stainless steel
Integral frost protection
SEDBUK rating A

Remove link SL–SL if wiring external voltage free controls.
LO – lock out indicator – optional.
RUN – run monitor – optional.

Malvern 30–70 condensing

Wall mounted

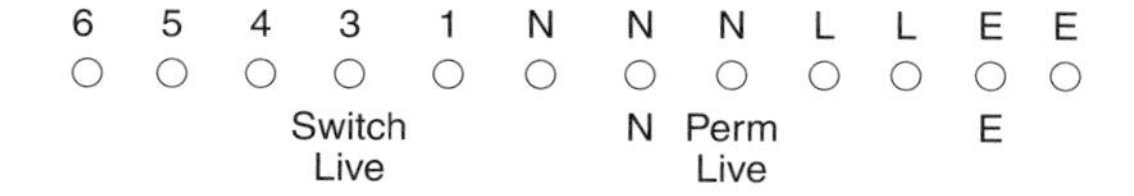

Fully pumped systems only
Suitable for use on sealed systems
Heat exchanger material, copper, aluminium
SEDBUK rating C

Maxol EM25

Wall mounted

L	E	N
Switch Live	E	N

Fully pumped systems only
Not suitable for use on sealed systems
Heat exchanger material, copper

Maxol EM40, EM50

Wall mounted

E	L	N	NP	LP	EP	1	2	3	4
E	L Mains	N		Pump		L	Switch Live	Internal Wiring	

Fully pumped systems only
Not suitable for use on sealed systems
Heat exchanger material, copper

Maxol Homewarm 600

Wall mounted

L	N	1	2	3	4	5	L	N	E
Timer Socket		Switch Live		L	N Pump	E	L	N Mains 240V	E

Terminals L–2–L are linked. Terminals N–4–N are linked

Fully pumped systems only
Not suitable for use on sealed systems
Heat exchanger material, copper

Maxol Microturbo 40

Wall mounted

E	1	2	3	4	5	L	N	E
E	Switch Live		L	N Pump	E	L	N Mains 240V	E

Terminals 2–L Mains are linked

Fully pumped systems only
Suitable for use on sealed systems
Heat exchanger material, copper
SEDBUK rating D

Maxol Morocco

Wall mounted

L	N	1	2	3	4	5	L	N	E
Timer Socket		Switch Live		L	N Pump	E	L	N Mains 240 V	E

Terminals L–2–L are linked. Terminals N–4–N are linked

Fully pumped systems only
Suitable for use on sealed systems
Heat exchanger material, copper
SEDBUK rating F

Maxol Mystique

Wall mounted

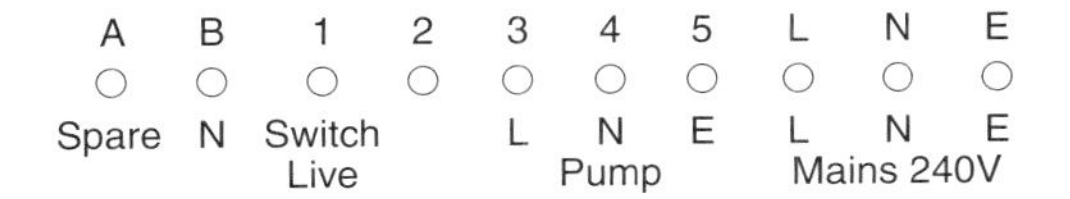

Terminals 2–L are linked

Fully pumped systems only
Not suitable for use on sealed systems
Heat exchanger material, copper

MHS Strata 1 condensing

Wall mounted

Connections of manufacturers control equipment and mains supply are via a plug-in system.

Fully pumped system only
Suitable for sealed systems only
Heat exchanger material, stainless steel
SEDBUK rating A

MHS Strata 2 condensing

Floor standing

Connections of manufacturers control equipment and mains supply are via a plug-in system.

Fully pumped system only
Suitable for sealed systems only
Heat exchanger material, stainless steel
SEDBUK rating A

MHS Strata 3 condensing

Floor standing

Connections of manufacturers control equipment and mains supply are via a plug-in system.

Fully pumped system only
Suitable for sealed systems only
Heat exchanger material, stainless steel
SEDBUK rating A

MHS Strata 38 condensing

Wall mounted

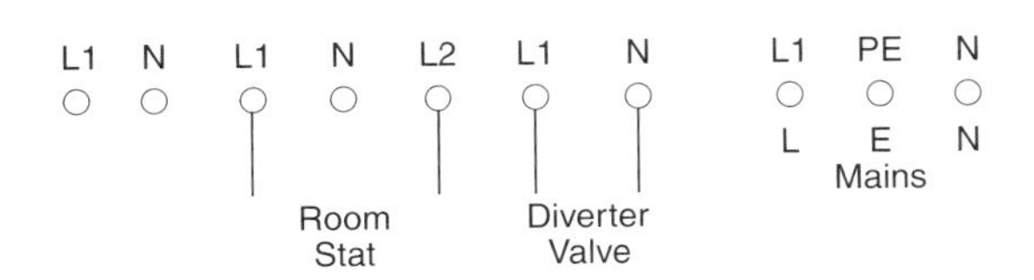

Fully pumped system only
Suitable for sealed systems only
Heat exchanger material, stainless steel
Integral frost protection
SEDBUK rating A

Wiring details are subject to the system into which the boiler is fitted. Brief details are given purely as a guide.

MHS Strata 38/46 condensing combination

Wall mounted

L1 N L1 N L2 L1 N L1 PE N

Room Stat

L E N
Mains

Fully pumped system only
Suitable for sealed systems only
Heat exchanger material, stainless steel
Integral frost protection
SEDBUK rating A

Wiring details are subject to the system into which the boiler is fitted. Brief details are given purely as a guide.

Myson Apollo

Wall mounted

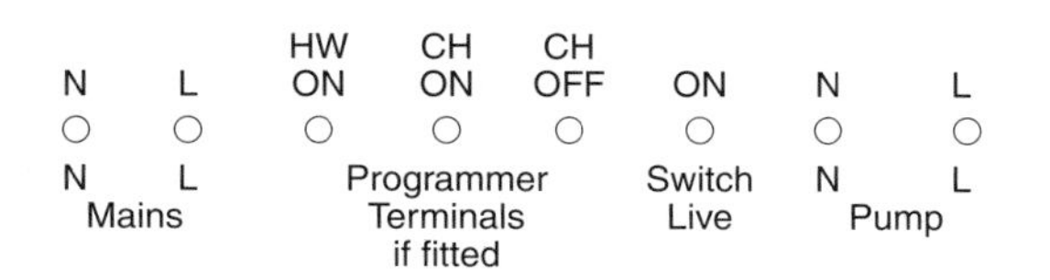

Fully pumped systems only
Suitable for use on sealed systems
Heat exchanger material, copper

Myson Housewarmer 2

Back boiler unit

L1 L N E

*Perm Live Switch Live N E

Not suitable for use on sealed systems
Heat exchanger material, cast iron

*Permanent live only required for fire front bulbs if fitted.

Myson Housewarmer electronic

Back boiler unit

L1 L N E Int

*Perm Live Switch Live N E

Not suitable for use on sealed systems
Heat exchanger material, cast iron

*Permanent live only required for fire front bulbs if fitted.

Myson Marathon 1500C

Floor standing

Remove link L–1 and 2–Pump L

Not suitable for use on sealed systems
Heat exchanger material, cast iron
Fit blue plug for gravity hot water
Fit red plug for fully pumped

Myson Midas B combination

Wall mounted

E2 N2 R1 R0 N L E

Room stat — Mains

Remove Link R1–R0 if wiring room thermostat

Suitable for sealed systems
Heat exchanger material, copper

Myson Midas Si combination

Wall mounted

L E N R1 R0 Int.

Mains 240V — Room stat 240V

Suitable for use on sealed systems
Heat exchanger material, copper

Connect room thermostat across terminals R1 and R0 after removing link. **Do not** connect room thermostat neutral. If wiring a frost thermostat, connect across terminals N and R1.

Myson Orion

Wall mounted

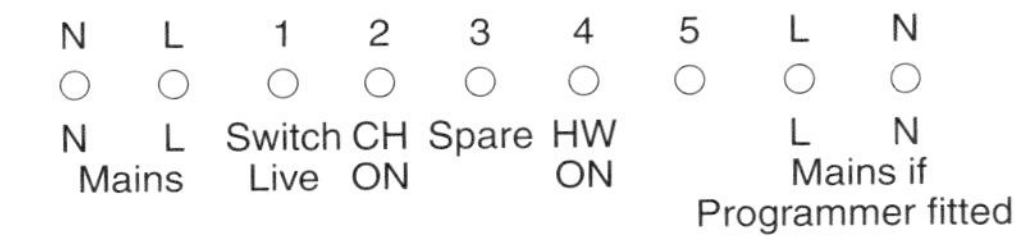

75 Si only suitable for use on sealed systems.
Others: suitable for use on sealed systems using optional kit.
Heat exchanger material, cast iron

Remove links L–2, 1–4, 4–5 and 2–4 if programmer fitted.

Ocean Alpha 240/280

Wall mounted

L N E 1 2

Mains — L In — L Out

Fully pumped systems only
Suitable for sealed systems
Heat exchanger material, copper/stainless steel

External controls such as time clock with voltage free contacts, or room thermostat suitable for use at 240V, should be connected between terminals 1 and 2 after removing link.

Ocean 80, OF, FF, FF style combination

Wall mounted

E L N 2 1

Mains 240V

Suitable for sealed systems
Heat exchanger material, copper

External controls such as time clock with voltage free contacts, or room thermostat suitable for use at 240V, should be connected between terminals 1 and 2 after removing link. Room thermostat neutral can be connected to terminal N.

Potterton Envoy

Wall mounted

L	N	SWL	E	PL	1	2	3
○	○	○	○	○	○	○	○
L	N	Switch Live	E	Pump Live	CH ON	HW ON	HW OFF
Mains							

Terminals 1, 2, 3, only used if optional timer used

Fully pumped systems only
Suitable for use on sealed systems
Heat exchanger material, aluminium
Integral frost protection
SEDBUK rating B

Potterton Fireside

Back boiler unit

L	N	E
○	○	○
Switch Live	N	E

Not suitable for use on sealed systems
Heat exchanger material, cast iron

Potterton Flamingo 2

Wall mounted

T	E	E	N	N	L	1	2	3
○	○	○	○	○	○	○	○	○
Int.		E	N		Switch Live	Spare		Spare

Terminals L–2 are linked

Not suitable for use on sealed systems
Heat exchanger material, cast iron

Potterton Flamingo 3

Wall mounted

Gravity hot water, pumped central heating

L	N	E	SWL	L	N	E
○	○	○	○	○	○	○
			Switch Live		N	E

Fully pumped systems

L	N	E	SWL	L	N	E
○	○	○	○	○	○	○
	Pump		Switch Live	Perm Live	N	E

Suitable for use on sealed systems
Heat exchanger material, cast iron

Potterton Housewarmer 45, 55

Back boiler unit

Gravity hot water, pumped central heating

N	L*	SWL	E	N	L
○	○	○	○	○	○
Mains		Switch Live	E		

* Permanent live only required if fire front has bulbs.

Heat exchanger material, cast iron

Fully pumped systems

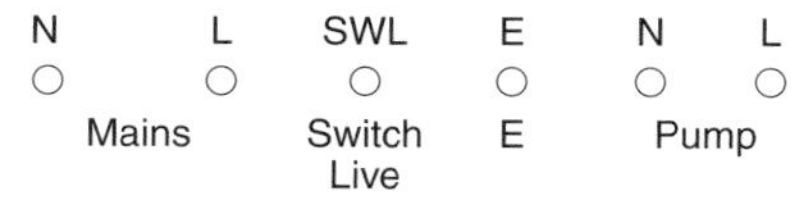

N	L	SWL	E	N	L
○	○	○	○	○	○
Mains		Switch Live	E	Pump	

Suitable for use on sealed systems when used on fully pumped systems

Potterton Kingfisher 2

Floor standing

Suitable for use on sealed systems using optional kit
Heat exchanger material, cast iron

Potterton Kingfisher Mf

Floor standing

Suitable for sealed systems only
Heat exchanger material, cast iron
SEDBUK rating D/E

Gravity hot water, pumped central heating

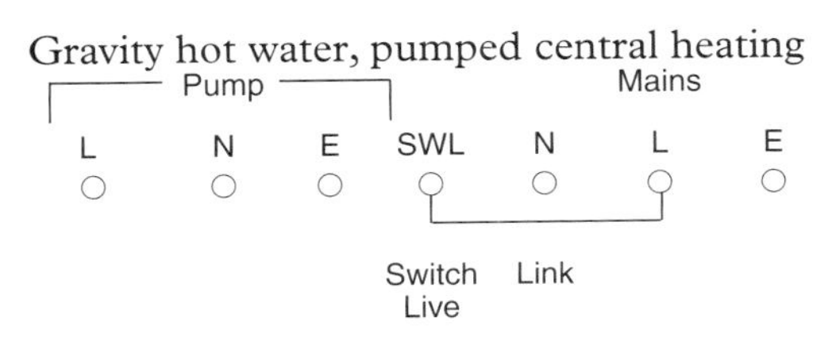

Fully pumped

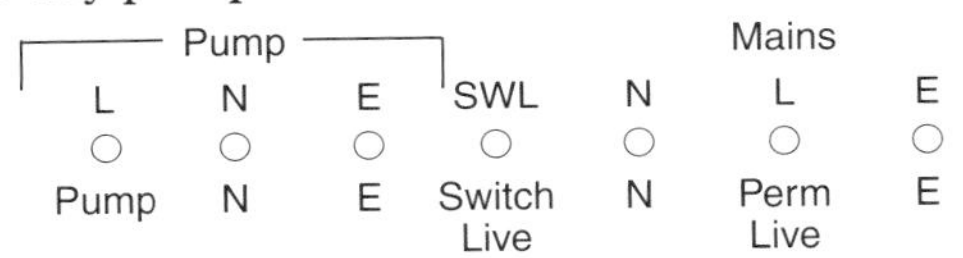

Note: On open vented, fully pumped systems, if no permanent live is available the boiler can be wired as follows providing that the system has a separate cold feed and vent pipe.

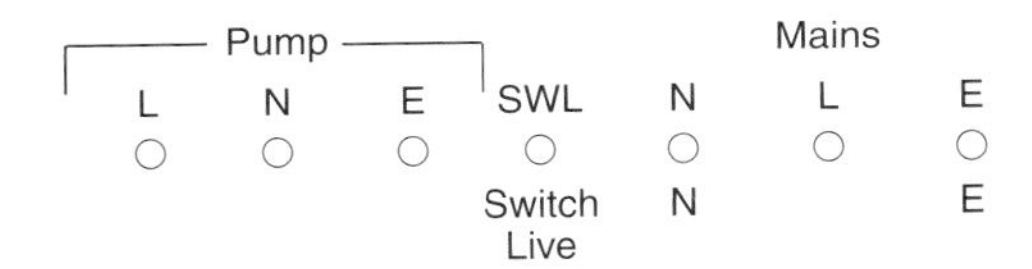

The overheat thermostat should be removed from the flow pipe and repositioned on the gravity return pipe using the bracket supplied.

Potterton Lynx combination

Wall mounted

Suitable for sealed systems
Heat exchanger material, copper

Potterton Myson Ultra System Boiler

Wall mounted

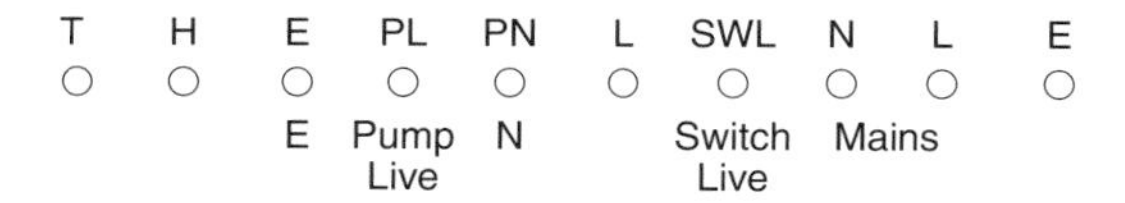

Suitable for sealed systems
Heat exchanger material, cast iron

The boiler has the facility to wire external room thermostat, cylinder thermostat and mid-position valve into the control box as follows:

Room stat Common	6	Cylinder stat Common	4	Mid-position valve:	Green–yellow	A
Room stat Demand	7	Cylinder stat Demand	ON		Blue	B
Room stat Neutral	2	Cylinder stat Satisfied	5		Grey	C
					Brown or white	D
					Orange	E

Potterton Osprey 2

Floor standing

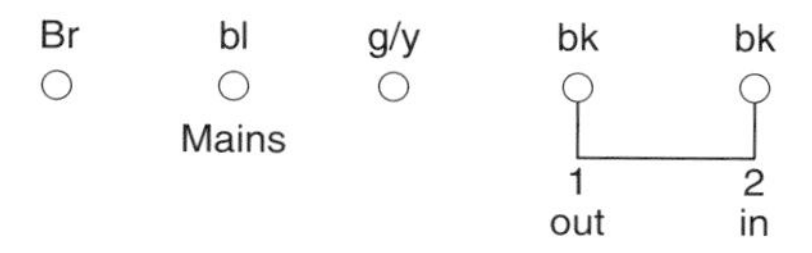

L & L are linked

Fully pumped system only
Suitable for sealed systems
Heat exchanger material, cast iron
SEDBUK rating D

Potterton Performa 24, 28, 28i combination

Wall mounted

Fully pumped system only
Suitable for sealed systems only
Heat exchanger material, copper
Integral frost protection, not model 24
SEDBUK rating B

External controls such as time clock with voltage free contacts or room thermostat suitable for use at 240V should be connected to terminal block after removing link. Connect room stat anticipator, if fitted, to N.

Potterton Powermax HE condensing

Floor Standing

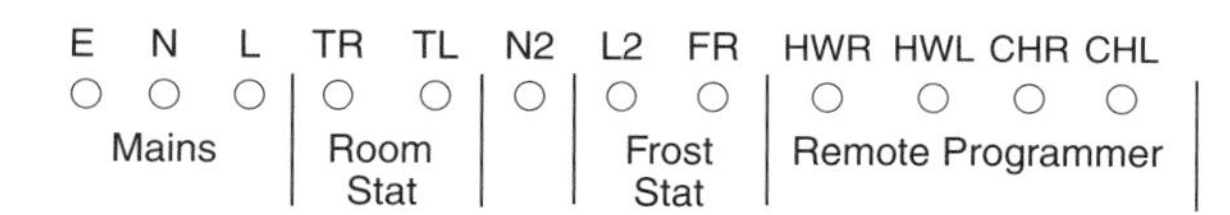

Fully pumped system only
Suitable for sealed systems only
Heat exchanger material, stainless steel
Integral frost protection
SEDBUK rating A

An integral programmer is supplied but if an external programmer is required the internal programmer must be switched off by moving the disable switch to the left.

Potterton Profile (also Netaheat Profile)

Wall mounted

Gravity hot water, pumped central heating

L	N	E	SWL	N	L
○	○	○	○	○	○
	Pump		Switch Live	N	*

Suitable for use on sealed systems
Heat exchanger material, cast iron
SEDBUK rating D/E/F

* From room thermostat demand or central heating 'on' if no room thermostat fitted.

Fully pumped

L	N	E	SWL	N	L
○	○	○	○	○	○
	Pump		Switch Live	N	Perm Live

Potterton Profile prima

Wall mounted

Gravity hot water, pumped central heating (not fanned flue model)

L	N	E	SWL	N	L
○	○	○	○	○	○
		E	Switch Live	N	

Fanned flue models: fully pumped systems only
Suitable for use on sealed systems
Heat exchanger material, cast iron

Fully pumped

L	N	E	SWL	N	L
○	○	○	○	○	○
	Pump		Switch Live	N	Perm Live

Potterton Promax 15HE, 24HE condensing

Wall mounted

S/L	E	N	P/F
○	○	○	○
Switch Live	E	N	Pump Live

Fully pumped system only
Suitable for sealed systems
Heat exchanger material, aluminium
SEDBUK rating A

The boiler does not require a permanent live. The pump should be connected to P/F if the system has TRV's throughout and no by-pass.

Potterton Promax System HE condensing

Wall mounted

S/L	E	N	P/F
○	○	○	○
Switch Live	E	N	

Fully pumped system only
Suitable for sealed systems only
Heat exchanger material, aluminium
SEDBUK rating A

Connection PF is an optional pump feed for use when fitting an additional external pump on a full TRV system with no by-pass fitted.

Potterton Puma combination

Wall mounted

External Timer — Room Stat

10	9	8	7	6	5	4	3	2	1
E	N	L	SWL	E	N	L	SWL	N	L
○	○	○	○	○	○	○	○	○	○
240V supply To Timer			Switch Live	E	N	Room Stat		Mains 240V	

Suitable for sealed systems
Heat exchanger material, copper
Integral frost protection
SEDBUK rating D

Potterton Suprima

Wall mounted

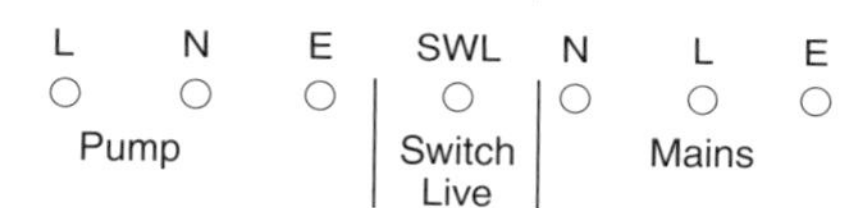

Fully pumped systems only
Suitable for use on sealed systems using optional kit
Heat exchanger material, cast iron
SEDBUK rating D/E

Potterton Suprima DV System

Wall mounted

This is a Suprima boiler with sealed system components and diverter valve added. It comes complete with a wiring centre to connect external controls.

Fully pumped system only
Suitable for sealed systems only
Heat exchanger material, cast iron
SEDBUK rating D

Potterton Suprima System & System POD

Wall mounted

This is a Suprima boiler with sealed system components and diverter valve added. Wiring as per Suprima. The 'Pod' version allows the sealed system components to be concealed behind a casing above the boiler.

Fully pumped system only
Suitable for sealed systems only
Heat exchanger material, cast iron
SEDBUK rating D

Potterton Tattler

Floor standing

E N L

E N Switch Live

Not suitable for use on sealed systems
Heat exchanger material, cast iron

Powermax 195 combination

Floor standing

The bulk of the water in the boiler's thermal store is used for central heating and also for heating the integral heat exchanger coil to supply domestic hot water. Model 185P has integral pump and 185CP has integral pump and programmer. Diagram shows basic model.

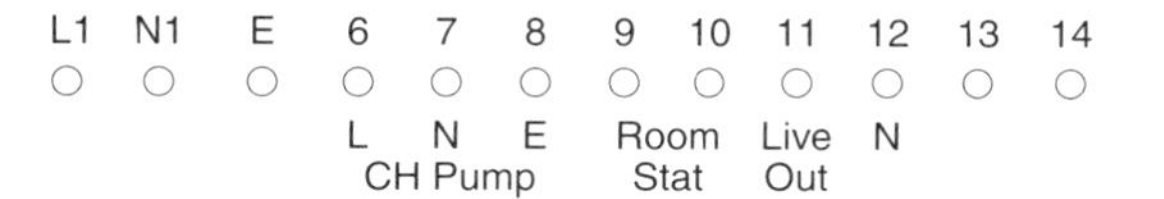

Not suitable for use on sealed systems
Heat exchanger material, copper

An internal programmer should have voltage free contacts. Link programmer live to hot water common. Connect the HW demand to terminal 14. Wire the central heating channel by removing link across 9–10 and wiring CH common and CH demand into 9–10. The programmer live should not be linked to the central heating switch.

Protherm 80E, 80EC, 100EC combination

Wall mounted

Fully pumped system only
Suitable for sealed systems only
Heat exchanger material, stainless steel
Integral frost protection
SEDBUK rating D

Radiant RCM, Comfort, RSF combination

Wall mounted

A 3-core lead is supplied for connection to an adjacent point. External controls such as time clock with voltage free contacts, or room thermostat, should be connected between terminals TA and OR after removing link. Room thermostat neutral should not be connected.

Suitable for sealed systems
Heat exchanger material, copper
SEDBUK rating D

Radiant R and RS Comfort heating only

Wall mounted

A 3-core lead is supplied for connection to an adjacent point. External controls such as time clock with voltage free contacts, or room thermostat, should be connected between terminals TA and OR after removing link. Room thermostat neutral should not be connected.

Suitable for sealed systems
Heat exchanger material, copper
SEDBUK rating D/E

Radiant RMA, RMAS combination

Incorporates 45 litre domestic hot water cylinder

A 3-core lead is supplied for connection to an adjacent point. External controls such as time clock with voltage free contacts, or room thermostat, should be connected between terminals TA and OR after removing link. Room thermostat neutral should not be connected.

Wall mounted

Suitable for sealed systems
Heat exchanger material, copper
SEDBUK rating D

Radiant Rain combination

Wall mounted

Fully pumped system only
Suitable for sealed systems only
Heat exchanger material, copper
Integral frost protection

Range Powermax combination

External Controls

24	L1	N1	E	25	N2	E	L2	26	27	28	N3
○	○	○	○	○	○	○	○	○	○	○	○
HW On		N	E		N	E	Prog Live			CH On	Prog N

Internal Programmer

24	L1	N1	E	25	N2	E	L2	26	27	28	N3
○	○	○	○	○	○	○	○	○	○	○	○
		N	E	L	N Pump	E			Live Out Room	Live In Stat	N

Connect mains supply to 3-way terminal block. Connect room stat between 27–28 after removing link.

Floor standing

Fully pumped system only
Only models SS suitable for sealed systems
SEDBUK rating C/D

Ravenheat 30B, 40B, 50B, 50S

E	N	L	D	C	N	7
○	○	○	○	○	○	○
E	N	Perm Live	Switch Live		N Pump	L Pump

When wiring external controls remove link D–C

Wall mounted

Fully pumped systems only
Suitable for use on sealed systems
(50S also incorporates a pressure vessel above the unit)
Heat exchanger material, copper

Ravenheat CF10/20, CF10/25 combination

Wall mounted

Suitable for sealed systems
Heat exchanger material, copper

External controls such as a time clock with voltage free contacts, or room thermostat suitable for use at 240V, should be connected between terminals C and D after removing link.

Ravenheat Combiplus combination

Wall mounted

L N C D
Mains 240V External controls

Suitable for sealed systems
Heat exchanger material, copper

External controls such as a time clock with voltage free contacts, or room thermostat suitable for use at 240V, should be connected between terminals C and D after removing link.

Ravenheat CSI 85, CSI 85T condensing combination

Wall mounted

Timer N L
Mains N L
Room Stat
Frost Stat

External controls must be voltage free

Fully pumped system only
Suitable for sealed systems only
Heat exchanger material, copper/aluminum
Integral frost protection
SEDBUK rating A

Ravenheat CS1 Primary condensing

Wall mounted

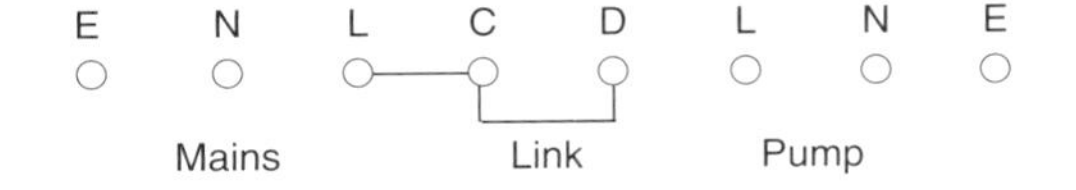

Fully pumped system only
Suitable for sealed systems
Heat exchanger material, copper/aluminium
SEDBUK rating A

Remove link C–D if fitting external controls. Terminal C is live out if required and terminal D is the control live in. The pump must be connected to the boiler terminal block.

Ravenheat CSI System condensing

Wall mounted

Timer N L
Mains N L
Room Stat
Frost Stat

External controls must be voltage free

Fully pumped system only
Suitable for sealed systems only
Heat exchanger material, copper/aluminum
Integral frost protection
SEDBUK rating A

Mid-position valve (Y plan)

E	E		VN	VL	V2	V3		CS1	CS2
○	○	○	○	○	○	○	○	○	○
E	E		blue	white	or	grey		cyl stat	

2 × 2-port motorized valves (S plan)

E	E		VN	VL	V2	V3		CS1	CS2
○	○	○	○	○	○	○	○	○	○
E	E		blue	brown CH valve	brown HW valve			cyl stat	

Ravenheat Little Star LS80, LS80T combination

Wall mounted

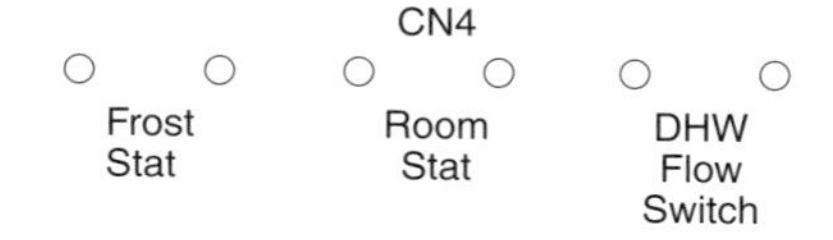

Fully pumped system only
Suitable for sealed systems only
Heat exchanger material, copper & stainless steel
Integral frost protection
SEDBUK rating D

A mains lead is supplied. A room thermostat should be connected as shown to PCB connection strip CN4 after removing link. External controls must be voltage free.

Ravenheat Merit

Wall mounted

Fully pumped systems only
Suitable for use on sealed systems
Heat exchanger material, copper

Ravenheat RSF 82E, RSF 82ET, RSF 84E, RSF 100E, combination RSF 84ET, RSF 100ET

Wall mounted

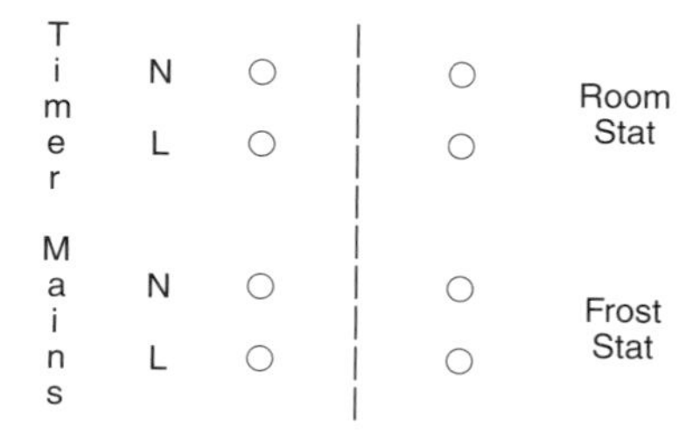

External controls must be voltage free

Fully pumped system only
Suitable for sealed systems only
Heat exchanger material, copper
Integral frost protection
SEDBUK rating D

Ravenheat RSF 820/20 combination

Wall mounted

L N C D
Mains 240V 24V

Suitable for sealed systems
Heat exchanger material, copper
SEDBUK rating E

External controls such as a time clock with voltage free contacts, or room thermostat suitable for use at 24V, should be connected between terminals C and D after removing link.

Rayburn GD 80 cooking appliance/boiler

Floor standing

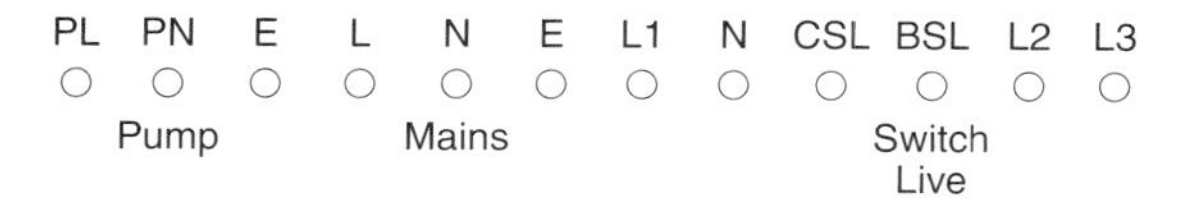

Fully pumped systems only
Suitable for use on sealed systems
Heat exchanger material, copper

If wiring external controls remove link L1–BSL. Pump Live must be wired into terminal PL.

Saunier Duval Ecosy 24E, 28E condensing combination

Wall mounted

Connect voltage free external controls via 3-way external controls plug after removing link. Do not connect room thermostat neutral.

Fully pumped system only
Suitable for sealed systems only
Heat exchanger material, aluminium
Integral frost protection
SEDBUK rating A

Saunier Duval Ecosy SB24E, SB28E condensing system

Wall mounted

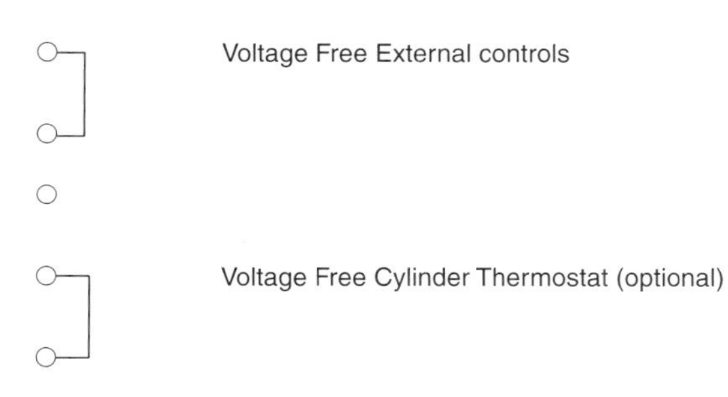

Fully pumped system only
Suitable for sealed systems only
Heat exchanger material, aluminium
Integral frost protection
SEDBUK rating A

Connect voltage free external controls via 5-way external controls plug after removing link. Do not connect room thermostat neutral.

Saunier Duval SD30e combination

Wall mounted

L	N	E		R1	R2
○	○	○		○	○
	Mains			Out	In

Fully pumped system only
Suitable for sealed systems only
Heat exchanger material, copper
Integral frost protection
SEDBUK rating D

Voltage free room thermostat and/or external timer should be connected between R1–R2 after removing link.

Saunier Duval SD 123C, 123F SD 223C, 223F

Wall mounted

		1	2	3
L ○	N ○	○	○	○
	E ○	○	○	○
	Mains 240V	6	5	4

Fully pumped systems only
Suitable for sealed systems
Heat exchanger material, copper

External controls such as a time clock with voltage free contacts, or room thermostat, should be connected between terminals 1 and 2 after removing link. Room thermostat neutral can be connected to terminal 5.

Saunier Duval SD 135C, 135F SD 235C, 235F

Wall mounted

		1	2	3
L ○	N ○	○	○	○
	E ○	○	○	○
	Mains 240V	6	5	4

Fully pumped systems only
Suitable for sealed systems
Heat exchanger material, copper

External controls such as a time clock with voltage free contacts, or room thermostat, should be connected between terminals 1 and 2 after removing link. Room thermostat neutral can be connected to terminal 5.

Saunier Duval SD 620F combination

Wall mounted

		1	2	3
L ○	N ○	○	○	○
	E ○	○	○	○
	Mains 240V	6	5	4

Suitable for sealed systems
Heat exchanger material, copper

External controls such as a time clock with voltage free contacts, or room thermostat, should be connected between terminals 2 and 3 after removing link. Room thermostat neutral can be connected to terminal 5.

Saunier Duval SD 623 combination

Wall mounted

		1	2	3
L ○	N ○	○	○	○
	E ○	○	○	○
	Mains 240V	6	5	4

Suitable for sealed systems
Heat exchanger material, copper

External controls such as a time clock with voltage free contacts, or room thermostat, should be connected between terminals 1 and 2 after removing link. Room thermostat neutral can be connected to terminal 5.

Saunier Duval SD 625M combination

Wall mounted

		1	2		3
L ○	N ○	○	○	24V	○
	E ○	○	○		○
	Mains 240V	6	5		4

Suitable for sealed systems
Heat exchanger material, copper

External controls such as a time clock with voltage free contacts, or room thermostat suitable for use at 24V, should be connected between terminals 2 and 3 after removing link. Room thermostat neutral can be connected to terminal 1.

Saunier Duval Isofast F combination

Wall mounted

A mains lead is supplied. A voltage free room thermostat can be connected to terminal block E after removing link.

Fully pumped system only
Suitable for sealed systems only
Heat exchanger material, copper
Integral frost protection
SEDBUK rating C

Saunier Duval Isomax combination

Wall mounted

A mains lead is supplied. A voltage free room thermostat can be connected to terminal block E after removing link.

Fully pumped system only
Suitable for sealed systems only
Heat exchanger material, copper
Integral frost protection
SEDBUK rating D

Saunier Duval Master Twin combination

Floor standing

Supplied with mains lead.

Suitable for sealed systems
Heat exchanger material, copper
Integral frost protection

External controls such as a time clock with voltage free contacts, or room thermostat, should be connected into 3-pin connector after removing link between terminals 2 and 3 in plug. Room thermostat neutral should not be connected.

Saunier Duval Master Twin L85E combination

Wall mounted

A 24V room thermostat can be connected to terminals 2 and 3 of the 3-way plug after removing the link. For frost protection to function, boiler must be set to frost or sun.

Fully pumped system only
Suitable for sealed systems only
Heat exchanger material, copper
Integral frost protection

Saunier Duval Sylva FF 24E combination

Wall mounted

A 24V room thermostat or voltage free external controls should be connected between terminals 2 and 3; otherwise a link should be fitted.

Fully pumped system only
Suitable for sealed systems only
Heat exchanger material, copper
Integral frost protection
SEDBUK rating D

Saunier Duval System 400

Wall mounted

Electrical connection is via a 3-pin plug

Fully pumped systems only
Suitable for use on sealed systems
Heat exchanger material, copper

Saunier Duval SB 23 Thelia

Wall mounted

Supplied with mains lead.

Fully pumped systems only
Suitable for sealed systems
Heat exchanger material, copper

External controls such as a time clock with voltage free contacts, or room thermostat, should be connected into 5-pin connector between top two terminals after removing link. Room thermostat neutral should not be connected. A cylinder thermostat can be connected across the bottom two terminals.

Saunier Duval 23, 23E Thelia combination

Wall mounted

Supplied with mains lead.

Suitable for sealed systems
Heat exchanger material, copper

External controls such as a time clock with voltage free contacts, or room thermostat, should be connected into 3-pin connector after removing link between 2 and 3 in plug. Room thermostat neutral should not be connected.

Saunier Duval 30E Thelia combination

Wall mounted

L N E
Mains 240V
1 2 3
6 5 4

Suitable for sealed systems
Heat exchanger material, copper

External controls such as a time clock with voltage free contacts, or room thermostat, should be connected between terminals 2 and 3 after removing link. Room thermostat neutral can be connected to terminal 5.

Saunier Duval Thelia Twin combination

Wall mounted

1 2 3
L ○ N ○ ○ ○ ○
E ○ ○ ○ ○
Mains 240V 6 5 4

Suitable for sealed systems
Heat exchanger material, copper
Integral frost protection

External controls such as a time clock with voltage free contacts, or room thermostat, should be connected between terminals 2 and 3 after removing link. Room thermostat neutral must not be connected.

Saunier Duval Thelia Twin 28E combination

Wall mounted

Fully pumped system only
Suitable for sealed systems only
Heat exchanger material, copper
SEDBUK rating D

A 24V room thermostat or voltage free external controls should be connected between terminals 2 and 3; otherwise a link should be fitted.

Saunier Duval Thema F-23E combination

Wall mounted

Connect external controls with voltage free contacts to 2-way connector after removing link. Mains lead supplied.

Fully pumped systems only
Suitable for sealed systems only
Heat exchanger material, copper

Saunier Duval Thema F SB 18E, SB 23E

Wall mounted

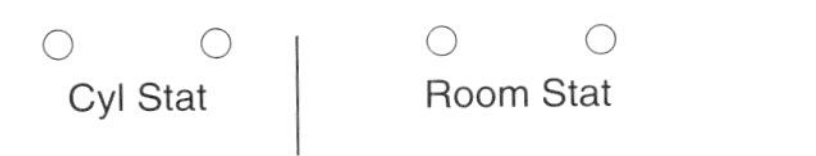

Fully pumped systems only
Suitable for sealed systems only
Heat exchanger material, copper

Saunier Duval 223 Themis combination

Wall mounted

1 2 3
L ○ N ○ ○ ○ ○
E ○ ○ ○ ○
Mains 240V 6 5 4

Suitable for sealed systems
Heat exchanger material, copper

For use without time switch or room thermostat, link together terminals 1, 2 and 3. With this wiring the control thermostat controls the operation of the burner – the pump runs continuously. To stop the pump working this way, disconnect the wiring loom connectors and insulate them from the earth of the boiler. With this layout the control thermostat controls the burner. The pump will stop when the burner goes out. For use with time switch and room thermostat, connect time switch voltage free contacts and room thermostat in series between terminals 1 and 2 and link together terminals 2 and 3. The neutral of the room thermostat must be connected to terminal 5.

Saunier Duval Thermaclassic F24E, F30E, Plus Condensing Combination

Wall mounted

Connection of external controls will be different depending on whether they are voltage free or mains voltage.
Voltage free – remove wire link from two-way connection on control box cover.
Mains voltage – refer to diagram below which shows the controls interface PCB.

○ 230v switched live from optional frost stat
○ 230v switched live from heating controls
○ Do not connect

If mains voltage external controls are used, the mains voltage heating controls plug should be installed on the controls interface PCB.

Fully pumped system only
Suitable for sealed systems only
Heat exchanger material, Stainless Steel
Integral frost protection
SEDBUK rating A

Saunier Duval Xeon FF

Wall mounted

Full Pumped

E	N	L	SL	9	PN	PL	E	K1	K2
E	N Mains	L	Switch Live		N	Pump Live	E		

Gravity Hot Water Pumped Central Heating

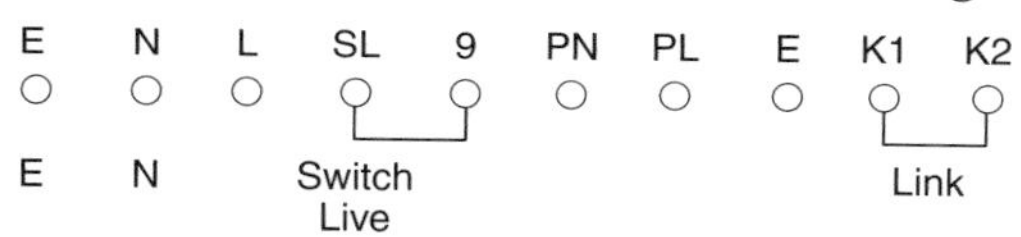

Link SL–9 on GHW/PCH Systems

Suitable for sealed systems
Heat exchanger material, cast iron
SEDBUK rating D/E

Sime Format 80C, 100C, 110C combination

Wall mounted

A room thermostat suitable for mains voltage should be connected to TA connector (10–11) after removing link.

Fully pumped system only
Suitable for sealed systems only
Heat exchanger material, aluminium
Integral frost protection
SEDBUK rating D

Sime Friendly, Friendly E combination

Wall mounted

A room thermostat should be connected to terminals 40–41 in the control box. In the Friendly model this should be suitable for mains voltage and in the Friendly E model this should be suitable for 24V.

Fully pumped system only
Suitable for sealed systems only
Heat exchanger material, aluminium
Integral frost protection
SEDBUK rating D

Sime Friendly Format combination

A room thermostat suitable for mains voltage should be connected to connector block TA, (22–23) after removing link.

Wall mounted

Fully pumped system only
Suitable for sealed systems only
Heat exchanger material, stainless steel
Integral frost protection
SEDBUK rating D

Sime Planet Dewy combination condensing

Mains lead supplied. A room thermostat should be connected to connector block J2, terminals 4–5.

Wall mounted

Fully pumped system only
Suitable for sealed systems only
Heat exchanger material, stainless steel
Integral frost protection
SEDBUK rating B

Sime Planet Super 4WM combination

Mains lead supplied. A voltage free room thermostat should be connected across TA–TA (5–6) after removing link. Comes with 'Logica' optimizer control.

Wall mounted

Fully pumped system only
Suitable for sealed systems
Heat exchanger material, aluminium
Integral frost protection

Sime Super 4 combination

A voltage free room thermostat should be connected to terminals 40–41 after removing link.

Floor standing (FS) and Wall mounted (WM)

Fully pumped system only
Suitable for sealed systems
Heat exchanger material, aluminium
Integral frost protection
SEDBUK rating D

Sime Super 90 combination

Connection via internal 5-pin plug

L	N	E	41	40
○	○	○	○	○
	Mains 240V		Room Stat 24V	

Wall mounted

Suitable for sealed systems
Heat exchanger material, copper
Frost thermostat and time clock integral to boiler.

Sime Super 90 & 105 MK11 combination

A room thermostat suitable for mains voltage should be connected to connector block TA, (22–23) after removing link.

Wall mounted

Fully pumped system only
Suitable for sealed systems only
Heat exchanger material, aluminium
Integral frost protection
SEDBUK rating D

Sime Superior MK 11

Wall mounted

2	1	E	N	L
○	○	○	○	○
Pump Live	Switch Live	E	N	Perm Live

Suitable for sealed systems only
Heat exchanger material, cast iron

For gravity HW. Pumped CH, the switch at the rear of the control box should be set to gravity.

Sinclair 40, 50

See Halstead 40H, 50H

Strebel SC 30 Condensing
SC 30C – Boiler
SC 30B – System
SC 30K – Combination

Wall mounted

6	5	4	3	2	1
○	○	○	○	○	○
				Room Stat	

Fully pumped system only
Suitable for sealed systems only
Heat exchanger material, stainless steel
SEDBUK rating A

External sensor and temperature sensor for cylinder may be fitted.

Strebel Super HR condensing

Wall mounted

Installation will normally include Brain room thermostat and controls. If these are to be replaced by other products they should be potential free.

Fully pumped system only
Suitable for sealed systems only
Heat exchanger material, stainless steel
Integral frost protection
SEDBUK rating A

Thermomatic RSM 15, 20, 25 combination

Wall mounted

14	13	12	11	E	N	L
○	○	○	○	○	○	○
					Mains	

Suitable for sealed systems
Heat exchanger material, copper

External controls such as a time clock with voltage free contacts, or room thermostat, should be connected between terminals 11–12 after removing link. Terminals 13–14 should remain linked.

Trianco Homeflame

L	N	E
○	○	○
Switch Live	N	E

Back boiler unit

Not suitable for use on sealed systems
Heat exchanger material, cast iron

Trianco Triancogas F

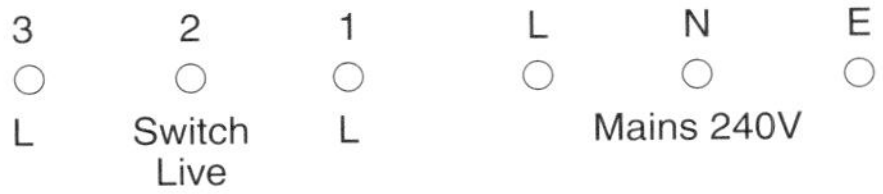

3	2	1	L	N	E
○	○	○	○	○	○
L	Switch Live	L		Mains 240V	

Wall mounted

Fully pumped systems only
Suitable for use on sealed systems
Heat exchanger material, copper

Trianco Triancogas RS

E	N	L1
○	○	○
E	N	Switch Live

Wall mounted

Fully pumped systems only
Suitable for use on sealed systems
Heat exchanger material, copper

Trianco Tristar

E	N	L	EP	NP	LP	1	2
○	○	○	○	○	○	○	○
E	N	L	E	N	L		Switch Live
	Mains			Pump			

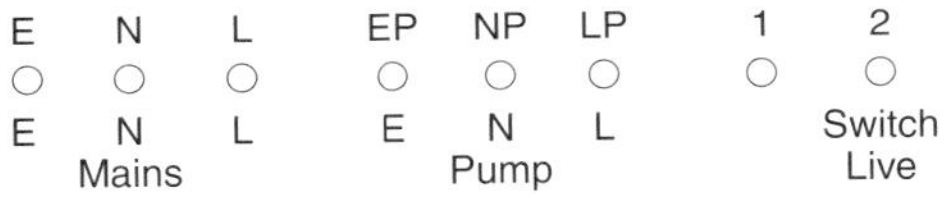

Terminal 1 can be used as live to external controls

Wall mounted

Fully pumped systems only
Suitable for use on sealed systems
Heat exchanger material, copper

Trianco Tristar Optima condensing combination

Wall mounted

Fully pumped system only
Suitable for sealed systems only
Heat exchanger material, aluminium
Integral frost protection
SEDBUK rating A

Trianco Tristar Optima condensing system

Wall mounted

Fully pumped system only
Suitable for sealed systems only
Heat exchanger material, aluminium
Integral frost protection
SEDBUK rating A

Trisave FS 60, 80, 80C condensing

Floor standing

A 4-core lead is supplied which should be connected as follows:

Grn yllw	Earth
Blue	Neutral
Black	Perm live
Brown	Switch live

Fully pumped systems only
Suitable for use on sealed systems using optional kit
Heat exchanger material, aluminium
The pump is connected into the boiler via 3-pin plug supplied.

Trisave Turbo 22, 30, 45, 60 condensing

Wall mounted

A 4-core lead is supplied which should be connected as follows:

Grn yllw	Earth
Blue	Neutral
Black	Perm live
Brown	Switch live

Fully pumped systems only
Suitable for use on sealed systems using optional kit
Heat exchanger material, aluminium
The pump is connected into the boiler via 3-pin plug supplied.

Vaillant, general information

Model codes:

VC = Vaillant Central (heating)
VCW = Vaillant Central (heating) Water, e.g. Combi
GB = Great Britain – following models may include GB but wiring is the same.
First two numbers denote kilowatt output and third number denotes flue type, i.e.
0 = Conventional, 1 = Balanced, 2 = Fanned
Following letters denote thus: E = Electronic, H = Natural gas, B = bottled gas.
Therefore a VCW 242E is a 24kW fanned flue electronic combination boiler.

Vaillant Ecomax 600 System condensing

Wall mounted

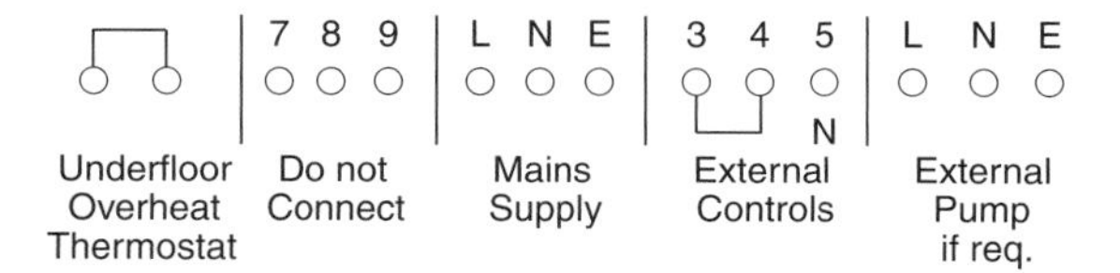

Fully pumped system only
Suitable for sealed systems only
Heat exchanger material, stainless steel
Integral frost protection
SEDBUK rating A

A room thermostat suitable for 230V and/or timer with voltage free controls should be connected across 3 and 4 after removing link. Terminal 5 can be used as a neutral connection.

Vaillant Ecomax 800 System combination condensing

Wall mounted

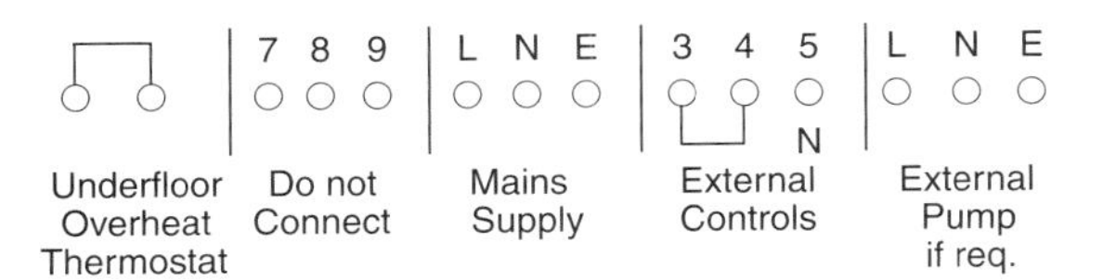

Fully pumped system only
Suitable for sealed systems only
Heat exchanger material, stainless steel
Integral frost protection
SEDBUK rating A/B

A room thermostat suitable for 230V and/or timer with voltage free controls should be connected across 3–4 after removing link. Terminal 5 can be used as a neutral connection.

Vaillant Ecomax pro condensing

Wall mounted

1 2 3 4 5 6 7 8

Perm Live | Switch Live | Frost Stat Live | N | E | E | N | Pump Live

Connect link 1–2 if no external controls

Fully pumped system only
Suitable for sealed systems
Heat exchanger material, stainless steel
Integral frost protection
SEDBUK rating A

Vaillant Thermocompact system

Wall mounted

7 8 9 L N E 3 4 5

Do Not Use | L N E Mains | N

Fully pumped system only
Suitable for sealed systems only
Heat exchanger material, stainless steel
Integral frost protection
SEDBUK rating D

A room thermostat suitable for 230V and/or timer with voltage free controls should be connected across 3–4 after removing link. Terminal 5 can be used as a neutral connection.

Vaillant Turbomax plus combination

Wall mounted

Fully pumped system only
Suitable for sealed systems only
Heat exchanger material, stainless steel
Integral frost protection
SEDBUK rating D

A room thermostat suitable for 230V and/or timer with voltage free controls should be connected across 3–4 after removing link. Terminal 5 can be used as a neutral connection.

Vaillant Turbomax Pro combination

Wall mounted

Fully pumped system only
Suitable for sealed systems only
Heat exchanger material, stainless steel
Integral frost protection
SEDBUK rating D

A room thermostat suitable for 230v and/or timer with voltage free controls should be connected across 3–4 after removing link. Terminal 5 can be used as a neutral connection.

Vaillant Turbomax Pro condensing

Wall mounted

1	2	3	4	5	6	7	8
Perm Live	Switch Live	Frost Stat Live	N	E	E	N	Pump Live

Connect link 1–2 if no external controls

Fully pumped system only
Suitable for sealed systems
Heat exchanger material, stainless steel
Integral frost protection
SEDBUK rating D/E

Vaillant VC 10, 15 TW3, 20 TW3

Wall mounted

6	5	4	3	2R	1MP
		External Switches		L	N
				Mains 240V	

Fully pumped systems only
Suitable for sealed systems
Heat exchanger material, copper

External controls such as a time clock with voltage free contacts, or room thermostat, should be connected between terminals 3 and 4 after removing link.

Vaillant VC 10-W, 15-W, 20-W

Wall mounted

6	5	4	3	2R	1MP
		External Switches		L	N
				Mains 240V	

Fully pumped systems only
Suitable for sealed systems
Heat exchanger material, copper

External controls such as a time clock with voltage free contacts, or room thermostat, should be connected between terminals 3 and 4 after removing link.

Vaillant VC 110, 180, 240 combination

Wall mounted

N	L	N 1	L 2	3	4	5
		Mains 240V				

Fully pumped systems only
Suitable for sealed systems
Heat exchanger material, copper

External controls such as a time clock with voltage free contacts, or room thermostat, should be connected between terminals 3 and 4 after removing link. Connect room thermostat neutral to terminal 5. Do not use terminals 7–12.

Vaillant VC 110 H, 180 H, 240 H

Wall mounted

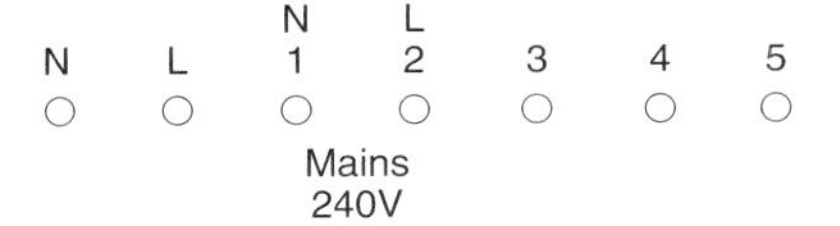

Fully pumped systems only
Suitable for sealed systems
Heat exchanger material, copper

External controls such as a time clock, or room thermostat, should be connected between terminals 3 and 4. Connect room thermostat neutral to terminal 5. Do not use terminals 7–12.

Vaillant VC 112 EH, 142 EH

Wall mounted

N L 1 2 3 4 5
N L
Mains
240V

Fully pumped systems only
Suitable for sealed systems
Heat exchanger material, copper

External controls such as a time clock, or room thermostat, should be connected between terminals 3 and 4. Connect room thermostat neutral to terminal 5. Do not use terminals 7–12.

Vaillant VC 112 E – RSF VC 142 E, 182 E, 242 E, 282 E

Wall mounted

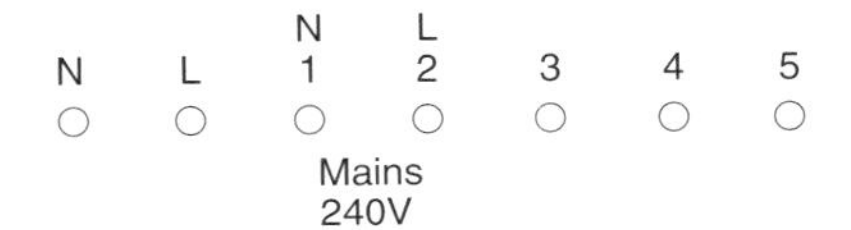

Fully pumped systems only
Suitable for sealed systems
Heat exchanger material, copper

External controls such as a time clock, or room thermostat, should be connected between terminals 3 and 4. Connect room thermostat neutral to terminal 5. Do not use terminals 7–12.

Vaillant VC 221 H

Wall mounted

N L 1 2 3 4 5
N L
Mains
240V

Fully pumped systems only
Suitable for sealed systems
Heat exchanger material, copper

External controls such as a time clock, or room thermostat, should be connected between terminals 3 and 4. Connect room thermostat neutral to terminal 5. Do not use terminals 7–12.

Vaillant VC Sine 18 W

Wall mounted

6 5 4 3 2R 1MP
External Switches
L N
Mains 240V

Fully pumped systems only
Suitable for sealed systems
Heat exchanger material, copper

External controls such as a time clock with voltage free contacts, or room thermostat, should be connected between terminals 3 and 4 after removing link.

Vaillant VCW 182 E, 242 E combination

Wall mounted

N	L	N 1	L 2	3	4	5
○	○	○	○	○	○	○
		Mains 240V				

Suitable for sealed systems
Heat exchanger material, copper

External controls such as a time clock, or room thermostat, should be connected between terminals 3 and 4. Connect room thermostat neutral to terminal 5. Do not use terminals 7–12.

Vaillant VC 221 combination

Wall mounted

N	L	N 1	L 2	3	4	5
○	○	○	○	○	○	○
		Mains 240V				

Fully pumped systems only
Suitable for sealed systems
Heat exchanger material, copper

External controls such as a time clock, or room thermostat, should be connected between terminals 3 and 4. Connect room thermostat neutral to terminal 5. Do not use terminals 7–12.

Vaillant VCW 240 H, 280 H combination

Wall mounted

N	L	N 1	L 2	3	4	5
○	○	○	○	○	○	○
		Mains 240V				

Suitable for sealed systems
Heat exchanger material, copper

External controls such as a time clock, or room thermostat, should be connected between terminal 3 and 4. Connect room thermostat neutral to terminal 5. Do not use terminals 7–12.

Vaillant VCW 242 EH, 282 EH combination

Wall mounted

N	L	N 1	L 2	3	4	5
○	○	○	○	○	○	○
		Mains 240V				

Suitable for sealed systems
Heat exchanger material, copper

External controls such as a time clock, or room thermostat, should be connected between terminals 3 and 4. Connect room thermostat neutral to terminal 5. Do not use terminals 7–12.

Vaillant VK-E, VKS-E

Floor standing

P E N L 3 4 5 6 7 8
Pump Live — Mains

Fully pumped systems only
Suitable for use on sealed systems
Heat exchanger material, cast iron

External controls such as a time clock with voltage free contacts, or room thermostat, should be connected between terminals 3 and 4 after removing link.

Vaillant VU 186 EH
VU 226 EH condensing, system

Wall mounted

E N L 3 4 5 7 8 9
E N L Mains — Do Not Use

Fully pumped systems only
Suitable for use on sealed systems
Heat exchanger material, copper
SEDBUK rating B

External controls such as a time clock with voltage free contacts, or room thermostat, should be connected between terminals 3 and 4 after removing link. When used in conjunction with a 3-port or 2 × 2-port motorized valve, the switched live can be connected into terminal 4.

Vaillant VUW 236 EH
VUM 286 EH
condensing, combination

Wall mounted

E N L 3 4 5 7 8 9
E N L Mains — Do Not Use

Suitable for sealed systems
Heat exchanger material, copper

External controls such as a time clock with voltage free contacts, or room thermostat, should be connected between terminals 3 and 4 after removing link.

Vokera 12–48 RS Mynute
12–48 RSE Mynute

Wall mounted

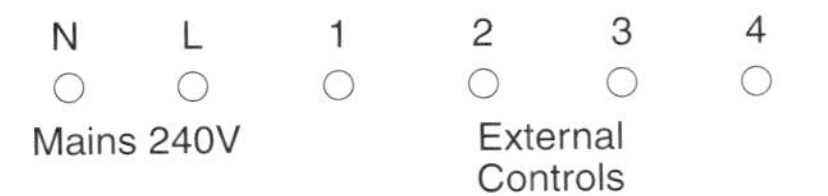

Suitable for sealed systems
Heat exchanger material, copper

Connect external controls such as a time clock, or room thermostat between terminals 2 and 3 after removing link. If wiring into a conventional fully pumped system, e.g. Y Plan, the switch live should be connected to terminal 3 after removing link.

Vokera 18–72 DMCF
21–84 DMCF combination

Wall mounted

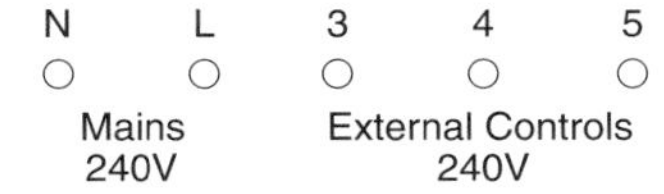

N	L	3	4	5
○	○	○	○	○
Mains 240V		External Controls 240V		

Suitable for sealed systems
Heat exchanger material, copper

Connect external controls such as a time clock, or room thermostat, between terminals 3 and 4 after removing link. Connect room thermostat neutral into terminal 5.

Vokera 20–80 flowmatic combination

Wall mounted

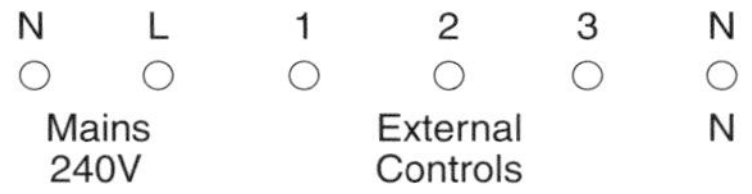

N	L	1	2	3	N
○	○	○	○	○	○
Mains 240V			External Controls		N

Suitable for sealed systems
Heat exchanger material, copper

Connect external controls such as a time clock, or room thermostat, between terminals 1 and 3.

Vokera 20–80 RS turbo combination

Wall mounted

N	L	3	4	5
○	○	○	○	○
Mains 240V		External Controls 240V		

Suitable for sealed systems
Heat exchanger material, copper

Connect external controls such as a time clock, or room thermostat, between terminals 3 and 4 after removing link. Connect room thermostat neutral into terminal 5.

Vokera 21–84 Turbo combination

Wall mounted

E	N	L	4	5
○	○	○	○	○
	Mains 240V		External Controls	

Suitable for sealed systems
Heat exchanger material, copper

Connect external controls such as a time clock with voltage free contacts, or room thermostat, between terminals 4 and 5 after removing link.

Vokera 21–84 DC turbo combination

Wall mounted

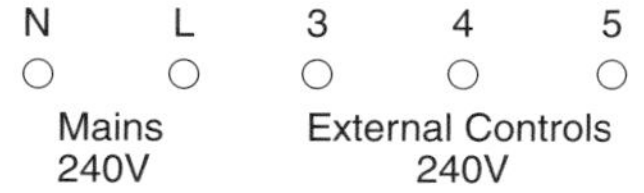

N	L	3	4	5
○	○	○	○	○
Mains 240V		External Controls 240V		

Suitable for sealed systems
Heat exchanger material, copper

Connect external controls such as a time clock, or room thermostat, between terminals 3–4 after removing link. Connect room thermostat neutral into terminal 5.

Vokera 20–80 Flowmatic combination

Wall mounted

N L 1 2 3 N

Mains 240V — External Controls — N

Suitable for sealed systems
Heat exchanger material, copper

Connect external controls such as a time clock, or room thermostat, between terminals 1 and 3.

Vokera Compact 24, 28 combinations

Wall mounted

Fully pumped system only
Suitable for sealed systems only
Heat exchanger material, copper
Integral frost protection
SEDBUK rating D

A 3-core mains lead is supplied. Connector strip is numbered 1–6 for clarity; it is not numbered in the boiler. L-4 is linked, N-6 is linked. Terminals 2 and 5 are spare for making connections. The timer clock motor should be wired from 4(L) and (6N). The control circuit for timer contacts and/or room thermostat are terminals 1 and 3.

Vokera Eclipse ESC condensing combination

Wall mounted

Integral timer. If connecting a room thermostat it must be suitable for 24V. The timer wire into terminal 2 of M4 should be connected to terminal 1 and room thermostat connected to 2 and 1.

External timer and/or room thermostat. Voltage free controls suitable for 24V should be connected between 2 and 3 on terminal block M4 after removing link.

Fully pumped system only
Suitable for sealed systems only
Heat exchanger material, aluminium
Integral frost protection
SEDBUK rating A

Vokera Eclipse ESS condensing system

Wall mounted

9 8 7 6 5 4 3 2 1

E L N

Fully pumped system only
Suitable for sealed systems only
Heat exchanger material, aluminium
Integral frost protection
SEDBUK rating A

External voltage-free controls suitable for 24V should be connected between terminals 6 and 4 after removing link.

Vokera Excell 80SP, 80E, 96E combination

Wall mounted

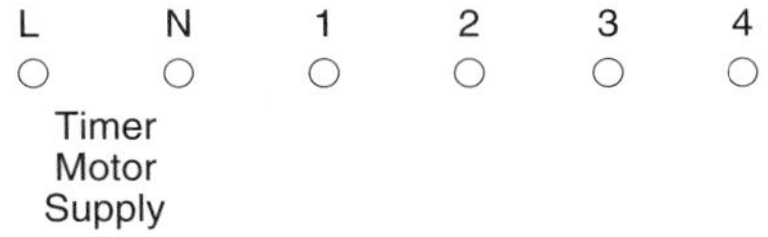

L	N	1	2	3	4
○	○	○	○	○	○
Timer Motor Supply					

Fully pumped system only
Suitable for sealed systems only
Heat exchanger material, copper

Connect voltage free controls to terminals 1–3 after removing link. Terminal 2 is spare and can be utilised.

Vokera Hydra condensing combination

Wall mounted

A mains supply should be connected to L + N on terminal block M2. A 24V room thermostat or voltage free control should be connected to terminal 2 and 3 on terminal block M7.

Fully pumped system only
Suitable for sealed systems only
Heat exchanger material, aluminium
Integral frost protection
SEDBUK rating A

Vokera Linea 7 combination

Wall mounted

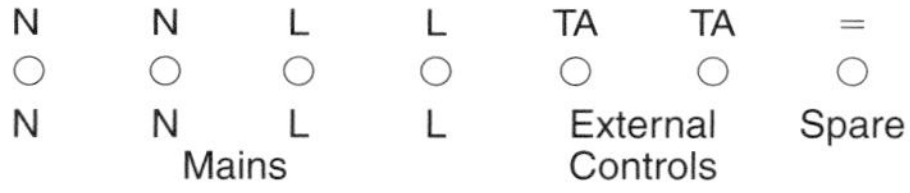

N	N	L	L	TA	TA	=
○	○	○	○	○	○	○
N	N	L	L	External		Spare
	Mains			Controls		

Fully pumped system only
Suitable for sealed systems only
Heat exchanger material, copper
Integral frost protection
SEDBUK rating D

It can be wired in conjunction with a cylinder and external controls, e.g. Honeywell 'S' Plan. Connect external controls across TA–TA after removing link. Connect room thermostat neutral to N. Terminal = is a spare terminal used to connect controls. This boiler can be used with 2 × 2-port motorised valves but not 1 × 3-way mid-position motorised valve.

Vokera Linea 24, 28 Plus combination

Wall mounted

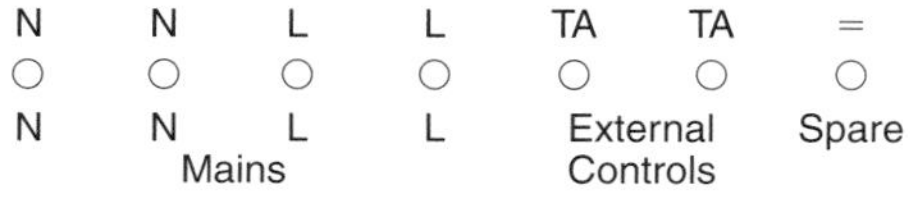

N	N	L	L	TA	TA	=
○	○	○	○	○	○	○
N	N	L	L	External		Spare
	Mains			Controls		

Fully pumped system only
Suitable for sealed systems only
Heat exchanger material, copper
Integral frost protection
SEDBUK rating D

It can be wired in conjunction with a cylinder and external controls, e.g. Honeywell 'S' or 'Y' Plan. Connect external controls across TA–TA after removing link. Connect room thermostat neutral to N. Terminal = is a spare terminal used to connect controls.

Vokera Linea Max combination

Floor standing

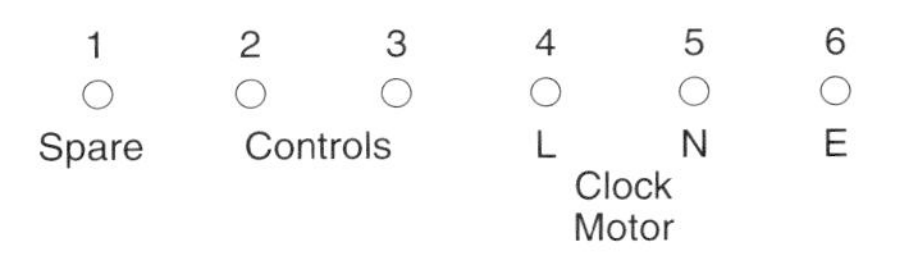

Fully pumped system only
Suitable for sealed systems only
Heat exchanger material, copper
Integral frost protection
SEDBUK rating D

Voltage free external controls and/or room thermostat should be connected to terminals 2–3 after removing link. Room thermostat neutral should be connected to 5 or N.

Vokera Meteor S90 system

Floor standing

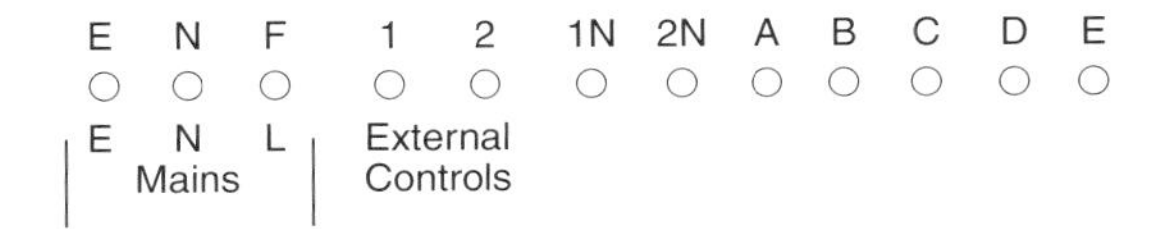

Fully pumped system only
Suitable for sealed systems only
Heat exchanger material, cast iron

Connect external controls between 1 and 2 after removing link.

Vokera Meteor V90

Floor standing

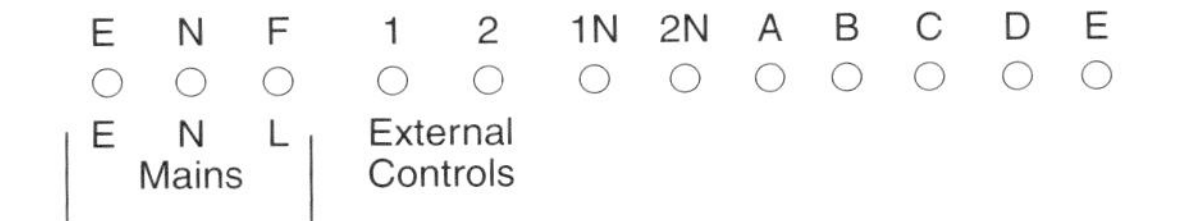

Fully pumped system only
Suitable for sealed systems
Heat exchanger material, cast iron

Connect external controls between 1 and 2 after removing link. Terminal 1 is switch live from external controls.

Vokera Mynute 10e, 14e, 20e system

Wall mounted

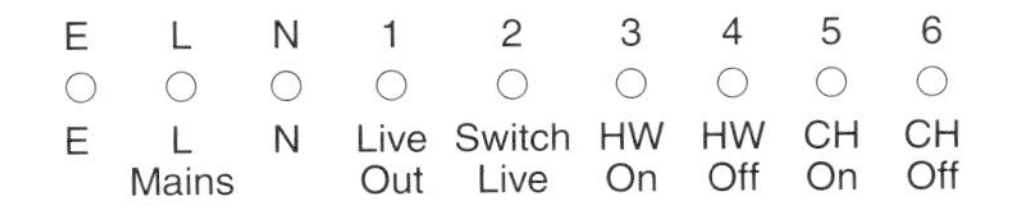

Fully pumped system only
Suitable for sealed systems only
Heat exchanger material, copper
SEDBUK rating D

Terminals 3–6 only apply if Vokera integral programmer fitted.

Vokera Mynute 10se, 14se, 28se system

Wall mounted

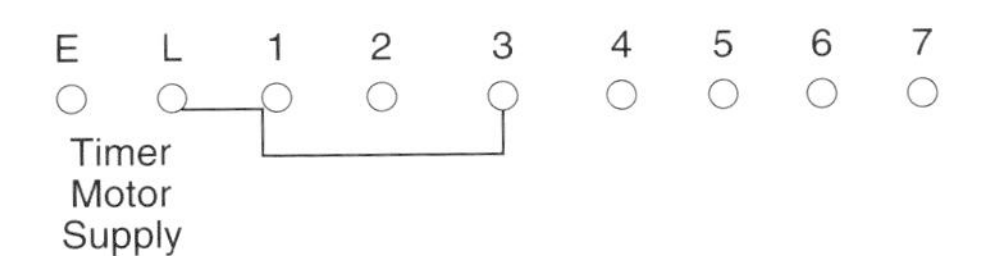

Fully pumped system only
Suitable for sealed systems only
Heat exchanger material, copper

A mains supply should be connected into the 3-way mains connector block. External controls suitable for mains voltage should be across terminals 1–3 after removing link.

Vokera Option combination

Wall mounted

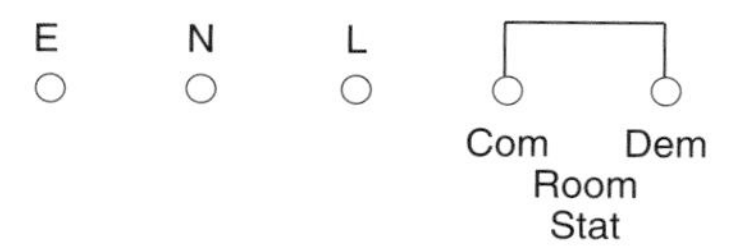

Fully pumped system only
Suitable for sealed systems only
Heat exchanger material, copper
Integral frost protection
SEDBUK rating D

The boiler has an integral time clock. An external timer and/or time clock should be wired into terminals as shown after removing link. The integral timer should be disconnected by removing black wire from terminal 5 of M3 and connecting to terminal 6.

Vokera Pinnacle 16, 26 condensing system

Wall mounted

A mains supply should be connected to L + N on terminal block M2. A 240V room thermostat or voltage free external controls such as motorized valve and switches, should be connected to terminals 2 and 3 on terminal block M7.

Fully pumped system only
Suitable for sealed systems only
Heat exchanger material, aluminium
Integral frost protection
SEDBUK rating A

Vokera Synergy 29 condensing system

Wall mounted

Fully pumped system only
Suitable for sealed systems only
Heat exchanger material, copper
Integral frost protection
SEDBUK rating B

Vokera Syntesi 25, 29, 35 condensing combinations

Wall mounted

Fully pumped system only
Suitable for sealed systems only
Heat exchanger material, aluminium/copper
Integral frost protection
SEDBUK rating B

Warmworld FFC 30/60, FFC 65/80 condensing

Wall mounted

LS L N E
Switch Live, Perm Live, N, E

E N L
E N L
Pump

Fully pumped system only
Suitable for sealed systems
Heat exchanger material, aluminum/copper/stainless steel
Integral frost protection
SEDBUK rating A

Warmworld HE condensing combination

Wall mounted

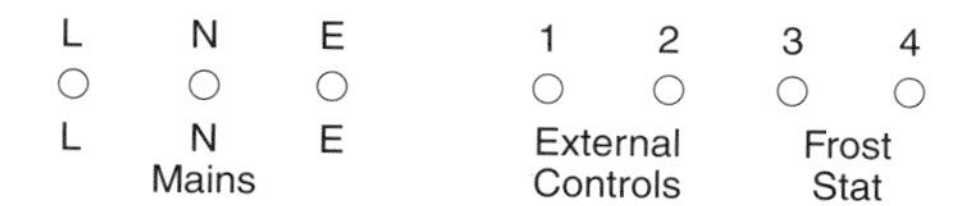

External controls must have voltage free contacts

Fully pumped system only
Suitable for sealed systems only
Heat exchanger material, aluminum/copper/stainless steel
SEDBUK rating A

Warm World HE30–HE70 condensing

Wall mounted

L	L	N	N	N	1	2	3	4	5	6
○	○	○	○	○	○	○	○	○	○	○
	L		N		Pump Live		Switch Live	HW Off	HW On	CH On

Integral Programmer if fitted (4, 5, 6)

Fully pumped systems only
Suitable for use on sealed systems
Heat exchanger material, aluminium and copper
SEDBUK rating C

Wickes 40, 50 MK2

See Halstead Blenheim

Wickes 42, 53

See Halstead 40H, 50H

Wickes 45F, 65F

See Halstead Balmoral

Wickes Combi

Wall mounted

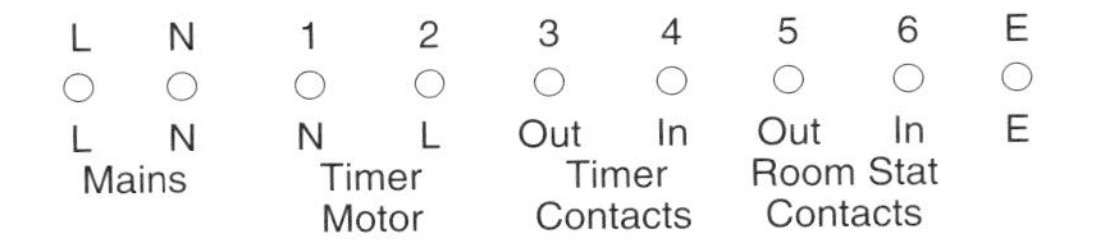

Suitable for sealed systems
Heat exchanger material, copper
SEDBUK rating D

If no timer or room stat Link 3–4 or 5–6 as appropriate.

Wickes Combi 30/90 combination.

Wall mounted

Connection via internal plug

L	N	E	41	40
○	○	○	○	○
	Mains		Room Stat 24V	

Suitable for sealed systems
Heat exchanger material, copper

If no room thermostat link 40–41. Time clock integral to boiler.

Worcester 9/14, 14/19, 19/24 CBi

Wall mounted

	○	Pump Live
	○	Pump Neutral
X1	○	Neutral
	○	Permanent Live
	○	Switch Live

Fully pumped system only
Suitable for sealed systems
Heat exchanger material, cast iron
Integral frost protection
SEDBUK rating D

Worcester 9.24 BF, OF electronic, combination

Wall mounted

X4 1 2 3 4 5
Room Stat N Frost Stat

Suitable for sealed systems
Heat exchanger material, copper

Worcester 9.24 BF, OF, MK 1.5 combination

Wall mounted

X4 1 2 3 4 5
Room Stat N Frost Stat

Suitable for sealed systems
Heat exchanger material, copper

Worcester 9.24 RSF 9.24 RSF 'E' electronic, combination

Wall mounted

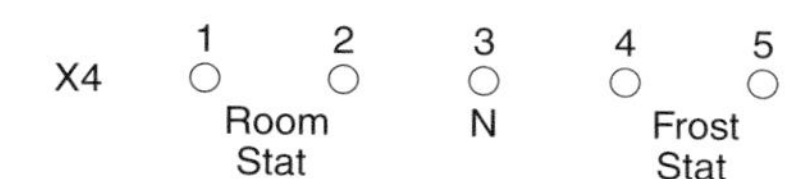

Suitable for sealed systems
Heat exchanger material, copper

Worcester 15Cbi, 24Cbi

Wall mounted

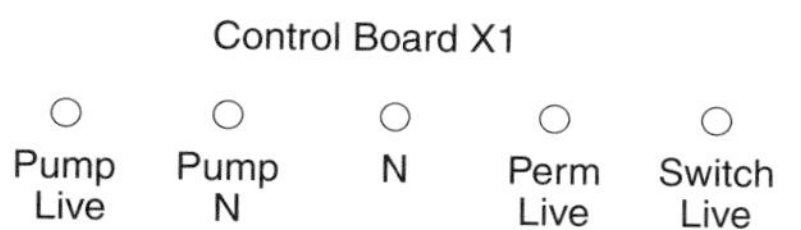

Suitable for sealed systems
Heat exchanger material, cast iron
Integral frost protection
SEDBUK rating E

Worcester 15Sbi, 24Sbi

Wall mounted

X2
LS LR N
Perm Live, Switch Live, N, Frost Stat

Remove link LS–LR if wiring external controls

Fully pumped system only
Suitable for sealed systems only
Heat exchanger material, copper
Integral frost protection
SEDBUK rating D

Worcester 24CDi combination

Wall mounted

ST12 ST8
L N NS LS LR SP
Mains

Suitable for sealed systems
Heat exchanger material, copper

External controls such as a time clock with voltage free contacts, or room thermostat, should be connected between terminals LS and LR after removing link. Room thermostat neutral can be connected to NS. If system uses wire free remote Digistat room thermostat – see ACL-Drayton Digistat RF1. If external frost protection is required also connect across LS–LR in addition to other wiring.

Worcester 24CDi, 28CDi, 35CDi, RSF combination

Wall mounted

To external controls – 230V

Fully pumped system only
Suitable for sealed systems only
Heat exchanger material, copper/stainless steel
Integral frost protection
SEDBUK rating D/E

Worcester 24i, 28i RSF combination

Wall mounted

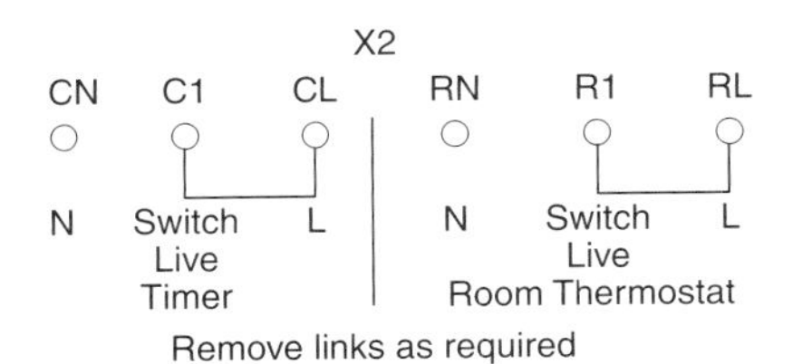

Fully pumped systems only
For sealed systems only
Heat exchanger material, copper
Integral frost protection
SEDBUK rating D/E

Worcester 25Si, 28Si, combination

Wall mounted

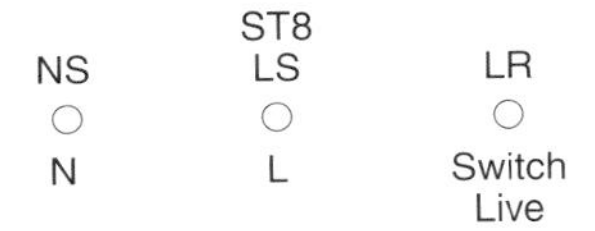

External controls – 230V

Fully pumped system only
Suitable for sealed systems only
Heat exchanger material, copper
SEDBUK rating D/E

Worcester 26CDi XTRA condensing combination

Wall mounted

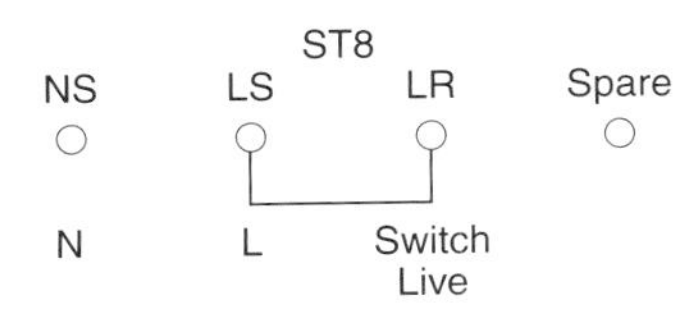

To external controls – 230V

Fully pumped system only
Suitable for sealed systems only
Heat exchanger material, copper
Integral frost protection
SEDBUK rating C

Worcester 240, BF, OF

Wall mounted

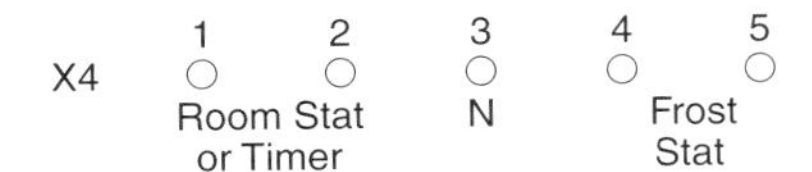

Suitable for sealed systems
Heat exchanger material, copper

External controls such as a time clock with voltage free contacts, or room thermostat suitable for use at 240V, should be connected as shown.

Worcester 240, BF, OF RSF

Wall mounted

X4 1 2 3 4 5
Room Stat or Timer, N, Frost Stat

Suitable for sealed systems
Heat exchanger material, copper

External controls such as a time clock with voltage free contacts, or room thermostat suitable or use at 240V, should be connected as shown.

Worcester 280, RSF

Wall mounted

X5 5 4 3 2 1
N, Room Stat or Timer, Frost Stat

Suitable for sealed systems
Heat exchanger material, copper

External controls such as a time clock with voltage free contacts, or room thermostat suitable for use at 240V, should be connected as shown.

Worcester 350 combination

Wall mounted

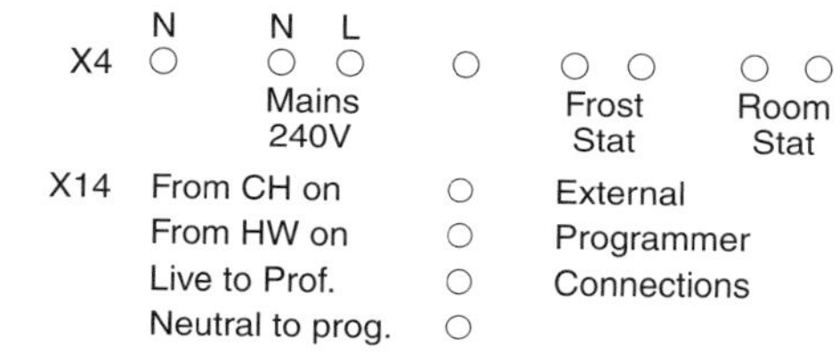

Suitable for sealed systems
Heat exchanger material, copper

Worcester Delglo 2 combination

Floor standing

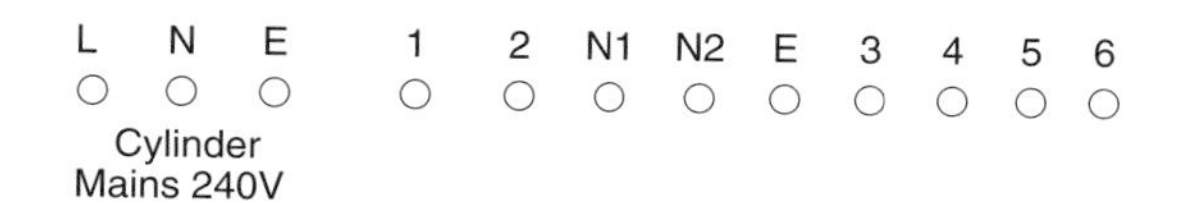

Suitable for use on sealed systems
Heat exchanger material, copper

External controls such as a time clock with voltage free terminals, or room thermostat, should be connected between terminals 2 and 3 after removing link. A frost thermostat should be of the double pole type wired common to terminal 6 and 2-pole connections to 1 and 2.

Worcester Delglo 3 combination

Floor standing

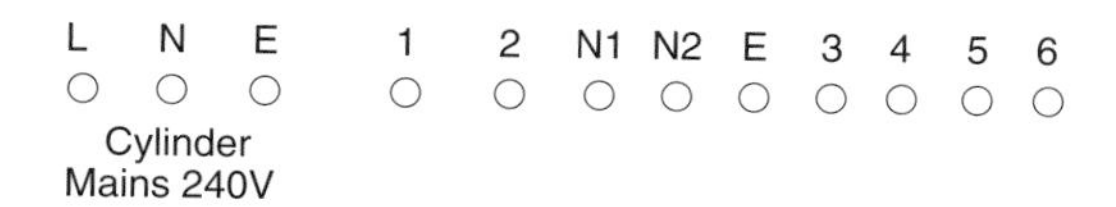

Suitable for use on sealed systems
Heat exchanger material, copper

External controls such as a time clock with voltage free terminals, or room thermostat, should be connected between terminals 2 and 3 after removing link. A frost thermostat should be of the double pole type wired common to terminal 6 and 2-pole connections to 1 and 2.

Worcester Greenstar HE combination condensing

Wall mounted

The boiler incorporates a plug-in facility for controls

Fully pumped system only
Suitable for sealed systems only
Heat exchanger material, aluminium
Integral frost protection
SEDBUK rating A

Worcester Greenstar HE system condensing

Wall mounted

The boiler incorporates a plug-in facility for controls

Fully pumped system only
Suitable for sealed systems only
Heat exchanger material, aluminium
Integral frost protection
SEDBUK rating A

Worcester Heatslave 9.24 BF, OF, MK 1 combination

Floor standing

L N E   E 1 2 3 N1 4 5 6
Mains 240V

Suitable for use on sealed systems
Heat exchanger material, copper

External controls such as a time clock with voltage free terminals, or room thermostat, should be connected between terminals 2 and 3 after removing link. Wire frost thermostat across terminals 4–5.

Worcester Heatslave 9.24 RSF combination

Wall mounted

L N E   1 2 3 N1 4 5 N 6
Mains 240V

Suitable for use on sealed systems
Heat exchanger material, copper

External controls such as a time clock with voltage free terminals, or room thermostat, should be connected between terminals 2 and 3 after removing link. Wire frost thermostat across main live and terminal 3.

Worcester Heatslave High Flow combination

Floor standing

Suitable for use on sealed systems
Heat exchanger material, copper

External controls such as a time clock with voltage free terminals, or room thermostat, should be connected between terminals 2 and 3 after removing link. A frost thermostat should be of the double pole type and connected common to terminal L and 2-pole connections to 1 and 3.

Worcester Heatslave Senior 6 combination

Floor standing

L N E N N E 1 2 3 4 5 6 7 8 9

Mains 240V

Suitable for use on sealed systems
Heat exchanger material, copper

External controls such as a time clock with voltage free terminals, or room thermostat, should be connected between terminals 2 and 4 after removing link. A frost thermostat should be of the double pole type and wired common to terminal 3 and 2-pole connections to 1 and 2.

Worcester High Flow 3.5 BF, OF, 4.5 BF, OF, 5.3 BF, OF combination

Floor standing

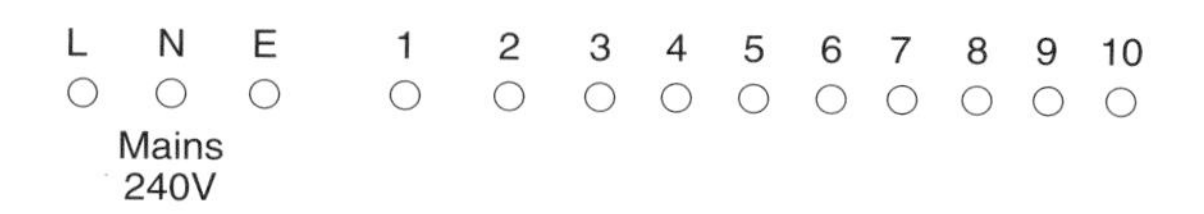

SS versions only, suitable for use on sealed systems
Heat exchanger material, copper

External controls such as a time clock with voltage free terminals, or room thermostat, should be connected between terminals 2 and 3 after removing link. A frost thermostat should be of the double pole type and connected common to terminal L and 2-pole connections to 1 and 3.

Worcester Highflow 400, OF, BF, RSF combination

Wall mounted

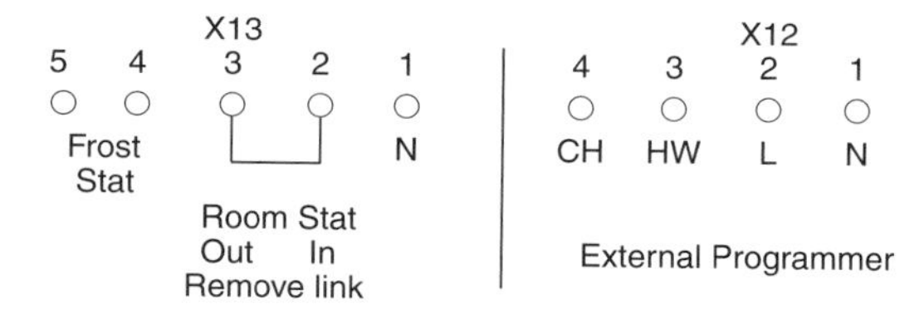

Fully pumped system only
Suitable for sealed systems only
Heat exchanger material, copper
SEDBUK rating D/E

Yorkpark Microstar 20 condensing

Wall mounted

Fully pumped systems only
Suitable for sealed systems
Heat exchanger material, aluminium

The above connections apply to a fully pumped system such as a Y plan. However, if only a room thermostat and/or time clock is to be used, then a permanent live should be wired to L and N, and the control wiring connected across TH–TH after removing link.

Yorkpark Microstar 20 condensing, combination

Wall mounted

TH TH N N L
N Switch Live

Suitable for sealed systems
Heat exchanger material, aluminium

The above connections apply to a fully pumped system such as a Y plan. However, if only a room thermostat and/or time clock is to be used, then a permanent live should be wired to L and N and the control wiring connected across TH–TH after removing link.

Yorkpark Microstar MC24G condensing

Floor standing

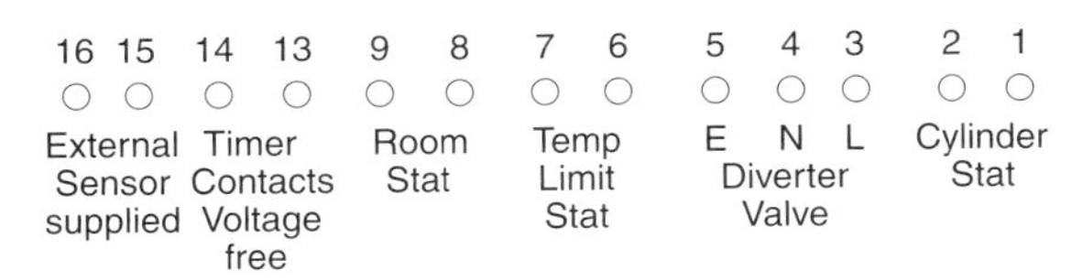

Fully pumped systems only
Suitable for sealed systems
Heat exchanger material, steel

Terminals 1, 2, 6, 7, 8, 9, 13, 14, 15, 16 are low voltage. A 240V mains should be connected into the separate terminal block.

Yorkpark Microstar MZ22C condensing

Wall mounted

TH TH N N L
N Switch Live

Fully pumped systems only
Suitable for sealed systems
Heat exchanger material, aluminium

The above connections apply to a fully pumped system such as a Y plan. However, if only a room thermostat and/or time clock is to be used, then a permanent live should be wired to L and N, and the control wiring connected across TH–TH after removing link.

Yorkpark Microstar MZ22S condensing, combination

Wall mounted

TH TH N N L
N l

Suitable for sealed systems
Heat exchanger material, aluminium

External controls such as a time clock with voltage free terminals, or room thermostat, should be connected between terminals TH–TH after removing link.

10

Boiler wiring – oil

Aquaflame Evolution

Floor standing

Fully pumped systems only
Suitable for sealed systems
Heat exchanger material, stainless steel
Integral frost protection
SEDBUK rating C

Aquaflame Eco-Avance 11 condensing

Floor standing

Fully pumped systems only
Suitable for sealed systems
Heat exchanger material, stainless steel
SEDBUK rating A

B. H. Associates Merlin 45/65, 65/95, 100/150, 150/185

Floor standing

Fully pumped systems only
Suitable for sealed systems
Heat exchanger material, stainless steel
Integral frost protection
SEDBUK rating C

B. H. Associates Merlin 2000 40/60, 60/80, 80/100

Floor standing

Fully pumped systems only
Suitable for sealed systems
Heat exchanger material, stainless steel
SEDBUK rating C

Boulter Bonus

1	2	3	4
○	○	○	○
Switch Live	N	Perm Live	E

Floor standing

Suitable for sealed systems
Heat exchanger material, mild steel
Integral frost protection
SEDBUK rating C

Boulter Camray 2.3

L	N	E
○	○	○
Switch Live	N	E

Camray 2 de-luxe has integral programmer

Floor mounted

Suitable for use on sealed systems
using optional kit
Heat exchanger material, mild steel

Boulter Camray 5 50/70 External

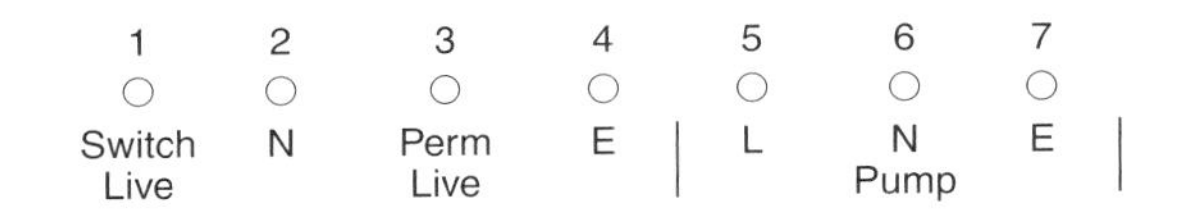

Wall mounted

Fully pumped system only
Suitable for sealed systems
Heat exchanger material, mild steel
Integral frost protection
SEDBUK rating C

Boulter Camray 5 50/70 Internal

Connection via 3-way terminal block

Wall mounted

Fully pumped system only
Suitable for sealed systems
Heat exchanger material, mild steel
SEDBUK rating C

Boulter Camray 5 150/200, 200/240

Connection via 3-way terminal block

Floor standing

Suitable for sealed systems
Heat exchanger material, mild steel
SEDBUK rating C

Boulter Camray 5 Combi

Connection is to L, N and E on terminal block

Floor standing

Fully pumped system only
Suitable for sealed systems
Heat exchanger material, mild steel
SEDBUK rating C

Boulter Camray 5 Combi External

Room and frost thermostats and external programmer should be connected into the boiler as shown. When fitting an external programmer the Summer/Winter switch will be inoperative as all wires should be cut from switch and connected as shown. Terminal 3 of JP12 is neutral for external programmer and/or room/frost thermostats. Mains connection is via a 3-pin plug.

Floor standing

Fully pumped system only
Suitable for sealed systems only
Heat exchanger material, mild steel
SEDBUK rating C

PCB JP9

1	2	3	4	5	6
○	○	○	○	○	○
CH On	Live Out		HW On		

PCB JP12

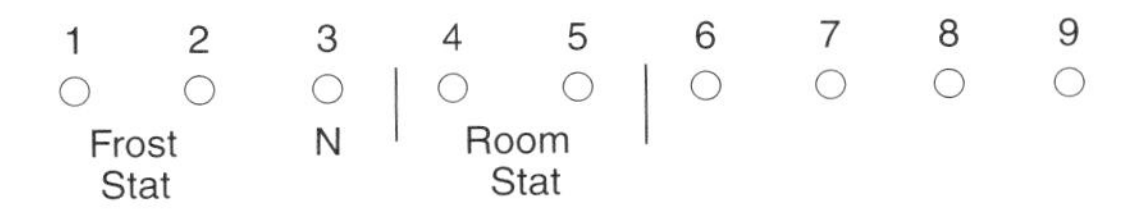

Boulter Camray 5 Kitchen & Utility Models

Floor standing

Connection via 3-pin plug.

Suitable for sealed systems
Heat exchanger material, mild steel
SEDBUK rating C

Boulter Camray 15/21

Wall mounted

3	2	1	E
○	○	○	
Perm Live	N	Switch Live	

Can be mounted externally with GRP case

Suitable for use on sealed systems
Heat exchanger material, mild steel
Integral frost protection

Boulter Camray Combi combination

Floor mounted

Mains via 3-pin plug. External controls, such as a time clock with voltage free contacts or room thermostat should be connected across terminals 3–5.

Suitable for use on sealed systems using optional kit
Heat exchanger material, mild steel

Boulter Camray Compact

Wall mounted

Connect to external controls via 3-pin plug supplied.

Fully pumped systems only
Suitable for use on sealed systems
Heat exchanger material, mild steel

Boulter Camray L60

Perm Live	Switch Live	N	E
○	○	○	○
Laux	L	N	E

Not suitable for use on sealed systems
Heat exchanger material, mild steel

Boulter Camray Pathfinder C, PJ

Floor mounted

L	N	E
○	○	○
Switch Live	N	E

Camray 2 de-luxe has integral programmer

Fully pumped system only
Suitable for use on sealed systems
Heat exchanger material, cast iron

Boulter Classic

Floor standing

Connection via 3-way terminal block.

Suitable for sealed systems
Heat exchanger material, mild steel
SEDBUK rating C

Boulter Eco-System

Floor standing

Connection via 3-way terminal block.

Fully pumped system only
Suitable for sealed systems only
Heat exchanger material, mild steel
SEDBUK rating C

Eco hometec EC condensing

Wall mounted

The integral pump is not the system pump. Unit supplied with Landis & Staefa Comfort Controller. A mains supply should be connected to the left-hand side of the integral wiring center. A voltage free control circuit, e.g. motorized valve and switches should be connected across terminals 13–14 after removing link.

Fully pumped system only
Suitable for sealed systems
Heat exchanger material, stainless steel
SEDBUK rating C

Eurocal Ambassador

Floor standing

Switch live should be connected to C of the high-limit stat and the neutral to the terminal connector.

Fully pumped system only
Suitable for sealed systems only
Heat exchanger material, mild steel
SEDBUK rating C

Eurocal Consul

Floor standing

Switch live should be connected to C of the high-limit stat and the neutral to the terminal connector.

Fully pumped system only
Suitable for sealed systems only
Heat exchanger material, mild steel
SEDBUK rating C

Eurocal Countryman

Floor standing

Switch live should be connected to C of the high-limit stat and the neutral to the terminal connector.

Fully pumped system only
Suitable for sealed systems only
Heat exchanger material, mild steel
SEDBUK rating C

Eurocal Countryman Combi

Floor standing

A room thermostat should be connected to terminals 3–4 on the 8 way connector strip after removing the link. A frost thermostat should be connected to terminals 5–3 on the 8-way connector strip. A thermostat neutral should be connected to terminal 6 on the 8-way connector strip.

Six-pin Plug

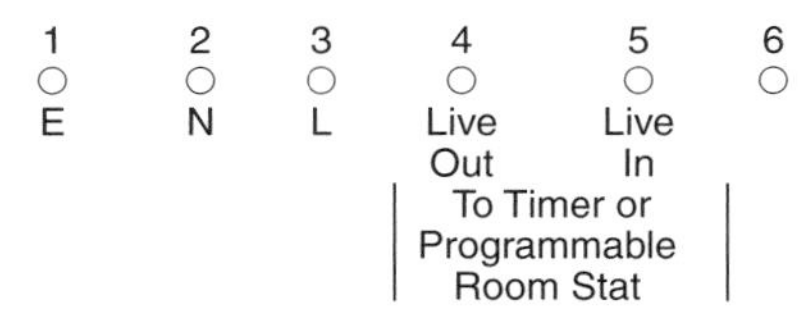

Fully pumped system only
Suitable for sealed systems only
Heat exchanger material, mild steel
SEDBUK rating C

Eurocal President

Floor standing

Connection is via a 3-pin plug.

Fully pumped system only
Suitable for sealed systems only
Heat exchanger material, mild steel
SEDBUK rating C

Eurocal Senator Combi

Floor standing

A room thermostat and/or timer should be connected to terminals 3–4 on the 8-way connector strip after removing the link. A frost thermostat should be connected to terminals 5–3 on the 8-way connector strip. A thermostat neutral should be connected to terminal 6 on the 8-way connector strip.

Fully pumped system only
Suitable for sealed systems only
Heat exchanger material, mild steel
SEDBUK rating C

Firebird Combi 90 combination

Floor standing

L1	E	N	L2
○	○	○	○
L	E	N	
	Mains		

A room thermostat can be connected to terminal strip after removing link

Fully pumped system only
Suitable for sealed systems only
Heat exchanger material, mild steel
Integral frost protection
SEDBUK rating C

Firebird Oylympic 'BH' Boilerhouse Model

Connect switched live to limit stat 'C'.
Connect Neutral to terminal block.

Floor standing

Fully pumped systems only
Suitable for sealed systems
Heat exchanger material, mild steel
Integral frost protection
SEDBUK rating C

Firebird Oylympic De Luxe

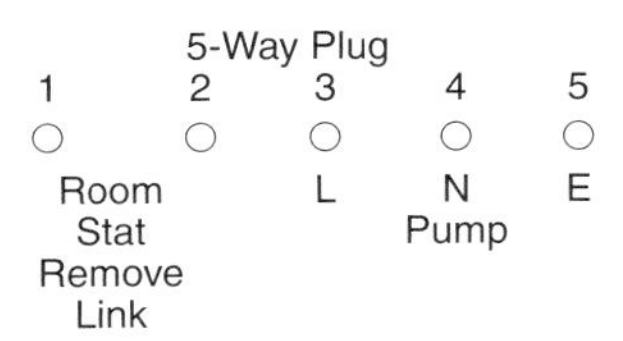

Connect mains to terminal strip

Floor standing

Fully pumped systems only
Suitable for sealed systems
Heat exchanger material, mild steel
Integral frost protection
SEDBUK rating C

Firebird Oylympic 'S' Standard

Connect switched live to limit stat 'C'.
Connect Neutral to terminal block.

Floor standing

Fully pumped systems only
Suitable for sealed systems
Heat exchanger material, mild steel
Integral frost protection
SEDBUK rating C

Firebird Oylympic system

Floor standing

Fully pumped systems only
Suitable for sealed systems
Heat exchanger material, mild steel
Integral frost protection
SEDBUK rating C

Gemini Triple Pass

L1	N2	E3	E4	N5	L6
○	○	○	○	○	○
Switch Live	N	E	E	N	L
			Burner supply		

Floor mounted

Suitable for use on sealed systems
Heat exchanger material, steel

Grant Combi 70 and 90 combination

Floor mounted

External controls such as time clock with voltage free contacts or room thermostat should be connected into time clock socket after removing red link.

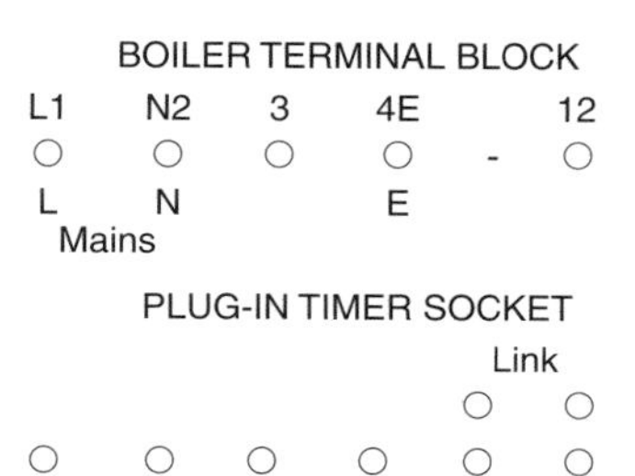

Suitable for use on sealed systems
Heat exchanger material, steel

Grant Combi MKII

Floor standing

Connect mains supply to L1, N2, E3. A room thermostat should be connected between terminals 7 and 12 after removing link. If fitting a timer remove link 13–14.

Fully pumped systems only
Suitable for sealed systems only
Heat exchanger material, stainless steel
SEDBUK rating C

Grant Combi 90 Outdoor

Floor standing

SL	SL	L	N	E	L	E	N	L			
1	2	3	4	5	6	7	8	9	10	11	12
CH Live	HW Live		Mains								

Fully pumped systems only
Suitable for sealed systems only
Heat exchanger material, stainless steel
Integral frost protection
SEDBUK rating C

If a time – switch is used link 2–3.
A room thermostat should be wired in series between the timer and boiler terminal SL1.

Grant Euroflame Boiler House

Floor standing

Connect switched live to C on the overheat thermostat.
Connect the neutral to burner neutral.

Suitable for sealed systems
Heat exchanger material, stainless steel
SEDBUK rating C

Grant Euroflame Kitchen/Utility

Floor standing

L	N	E	E	N	L
1	2	3	4	5	6
Switch Live	N	E		Burner	

Suitable for sealed systems
Heat exchanger material, stainless steel
SEDBUK rating C

Grant Euroflame Sealed system

L	N	E	L	N	E	E	N	L
1	2	3	4	5	6	7	8	9
SW Live	N	E		Pump			Burner	

Floor standing

Fully pumped systems only
Suitable for sealed systems
Heat exchanger material, stainless steel
SEDBUK rating C

Grant Euroflame Utility

L1	N2	N3	E4	N5	L6
Switch Live	N	E	E	N	L
				Burner supply	

Floor mounted

Suitable for use on sealed systems
Heat exchanger material, steel

Grant Multi Pass

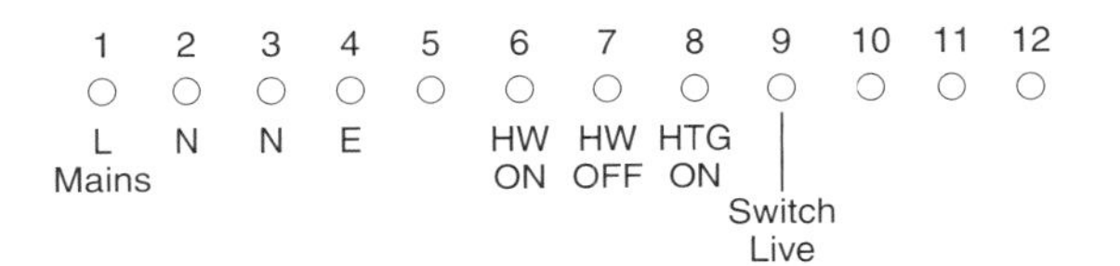

Floor mounted

Suitable for use on sealed systems
Heat exchanger material, steel

Grant Multi Pass Boiler House

Connect switched live to C on the overheat thermostat.
Connect the neutral to burner neutral.

Floor standing

Suitable for sealed systems
Heat exchanger material, stainless steel
SEDBUK rating C

Grant Multi Pass Kitchen 50/70, 70/90, 90/140

If fitting an integral programmer connect as follows: Live–1, Neutral–2, Earth–3, switched live-9 and remove link 1–9(L–BLR).
If fitting an external programme connect as follows: Switched Live-1, Neutral-2, Earth-3.

Floor standing

Suitable for sealed systems
Heat exchanger material, stainless steel
SEDBUK rating C

Grant Multi Pass system

1	2	3	4	5	6	7	8	9	10	11	12
L	N	N	E	Pump	HW On	HW Off	CH	BLR L	E	N	L
Mains			Pump							Burner	

Floor standing

Fully pumped systems only
Suitable for sealed systems only
Heat exchanger material, stainless steel
SEDBUK rating C

External controls with no internal programmer – connect switched live to 1 or 9.
Internal programmer with no external controls – remove link 1–9.
Internal programmer with external controls – remove link 1–9, 6–8–9, connect switched live to 9.

Grant Outdoor

Floor standing

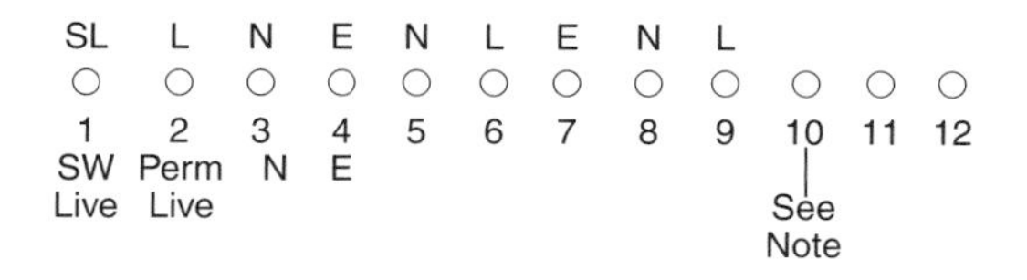

Suitable for sealed systems with kit
Heat exchanger material, stainless steel
Integral frost protection
SEDBUK rating C

Remove link 10–12 and connect room stat 'satisfied' on fully pumped systems.
The internal frost protection will not function unless a permanent live is connected to terminal L2.

Grant Vortex Kitchen condensing

Floor standing

If fitting an integral programmer, remove link 1–8(L–boiler feed), Connect as follows:
Live–1, Neutral–2, Earth–3, Switched Live–8
If fitting an external programmer connect as follows:
Switched Live-1, Neutral–2, Earth–3

Fully pumped system only
Suitable for sealed systems
Heat exchanger material, stainless steel
SEDBUK rating A

Grant Vortex Kitchen system condensing

Floor standing

Connect Mains to 1L, 2N, 3E
Connect Pump to 4L, 5N, 6E

Fully pumped system only
Suitable for sealed systems only
Heat exchanger material, stainless steel
SEDBUK rating A

Grant Vortex Utility condensing

Floor standing

L	N	E	E	N	L
1	2	3	4	5	6
SW Live	N	E		Burner	

Fully pumped systems only
Suitable for sealed systems
Heat exchanger material, stainless steel
SEDBUK rating A

Grant Vortex Utility system condensing

Floor standing

L	N	E	L	N	E	E	N	L
1	2	3	4	5	6	7	8	9
SW Live	N	E	L	N Pump	E		Burner	

Fully pumped systems only
Suitable for sealed systems
Heat exchanger material, stainless steel
SEDBUK rating A

Heating World Grandee Combi

Floor standing

If fitting a room stat, connect between CH-On and CH after removing existing connection.

Fully pumped system only
Suitable for sealed systems only
Heat exchanger material, mild steel
SEDBUK rating C

Heating World Grandee External

Fully pumped systems only
Suitable for sealed systems
Heat exchanger material, mild steel
Integral frost protection
SEDBUK rating C

Heating World Grandee Combi

Wall mounted

As Grandee Combi Floor.

Heating World Grandee Combi Compact

Wall mounted

As Grandee Combi Floor.

Heating World Sorrento

Back boiler unit

Connection is via a 3-pin plug.

Fully pumped systems only
Suitable for sealed systems
Heat exchanger material, stainless steel
SEDBUK rating C

HRM Starflow

Floor standing

N	E	L
○	○	○
N	E	Switch Live

Suitable for sealed systems
Heat exchanger material, mild steel
SEDBUK rating C

HRM Wallstar

Wall mounted

○	○	○	○
Switch Live	N	P-L	E

Integral frost thermostat
Suitable for use on sealed systems
Heat exchanger material, mild steel

HRM Wallstar combination

Wall mounted

Integral timer and mains lead fitted.
A room thermostat should be wired between 5A–6A after removing link.

Suitable for sealed systems
Heat exchanger material, mild steel
SEDBUK rating C

HRM Starflow

Floor mounted

SW-L	○
N	○
E	○

Suitable for use on sealed systems
Heat exchanger material, mild steel

Perrymatic Jetstreme MK3

CONTROL BOX				TERMINAL BLOCK				
1	2	3	4	5	6	7	8	9
○	○	○	○	○	○	○	○	○
		N		Switch Live*				

*via separate terminal block under burner cover

Floor mounted

Suitable for use on sealed systems
Heat exchanger material, steel

Potterton Statesman

Connection via 3-pin plug.

Floor standing

Fully pumped system only
Suitable for sealed systems
Heat exchanger material, steel
SEDBUK rating C

Potterton Statesman system

Mains connection via 3-pin plug. A room thermostat can be connected across L-S/L of the 6-way boiler terminal plug after removing link.

Floor standing

Fully pumped system only
Suitable for sealed systems
Heat exchanger material, steel
SEDBUK rating D

Potterton Statesman Flowsure combination

Mains connection via 3-pin plug. A room thermostat can be connected across the two middle terminals of the 6-way timer plug after removing link.

Floor standing

Fully pumped system only
Suitable for sealed systems
Heat exchanger material, steel
SEDBUK rating D

Potterton Statesman Flowsure & Storage combination

Mains connection via 3-pin plug. A room thermostat can be connected across the two middle terminals of the 6-way timer plug after removing link.

Floor standing

Fully pumped system only
Suitable for sealed systems
Heat exchanger material, steel
SEDBUK rating D

Thermecon Option

Connection via 3-way plug.

Floor standing

Suitable for sealed systems
Heat exchanger material, steel
SEDBUK rating C

Thermecon Option system

Floor standing

Connection via 3-way plug.

Suitable for sealed systems
Heat exchanger material, steel
SEDBUK rating C

Thermecon Select External

Floor standing or Wall mounted

1	E	N	2
○	○	○	○
Perm Live	E	N	Switch Live

Connection via 4-way plug.

Fully pumped systems only
Suitable for sealed systems
Heat exchanger material, steel
Integral frost protection
SEDBUK rating C

Thermecon Select Floor

Floor standing

Connection via 3-way plug.

Fully pumped systems only
Suitable for sealed systems
Heat exchanger material, steel
SEDBUK rating C

Thermecon Select Wall

Wall mounted

Connection via 3-way plug.

Fully pumped system only
Suitable for sealed systems
Heat exchanger material, steel
SEDBUK rating C

Thorn Panda

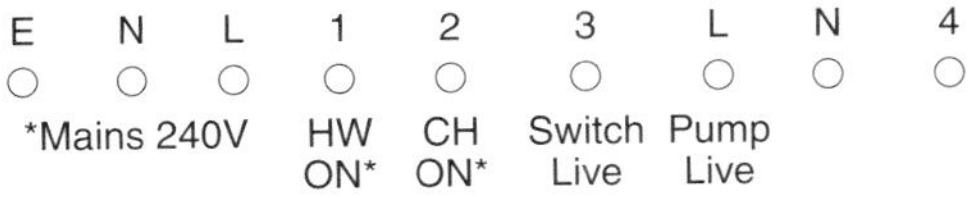

Floor mounted

E	N	L	1	2	3	L	N	4
○	○	○	○	○	○	○	○	○
*Mains 240V			HW ON*	CH ON*	Switch Live	Pump Live		

Not suitable for use on sealed systems
Heat exchanger material, steel

*If integral programmer fitted.
If used on fully pumped system remove links L–1–3, L–2. Link 3–pump live.

Trianco Centrajet 13/17

Wall mounted

E	N	L		
○	○	○	○	○
E	N	Switch Live		

Not suitable for use on sealed systems
Heat exchanger material, mild steel

Trianco Centramatic 13/17

Floor mounted

Not suitable for use on sealed systems
Heat exchanger material, mild steel

Without integral programmer

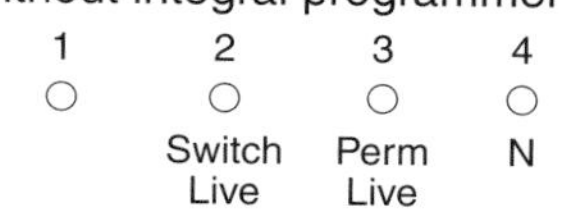

Remove Link 1–2

Trianco Centramatic 40, 55, 80

Floor mounted

Not suitable for use on sealed systems
Heat exchanger material, mild steel

2 way connector on 40 and 55

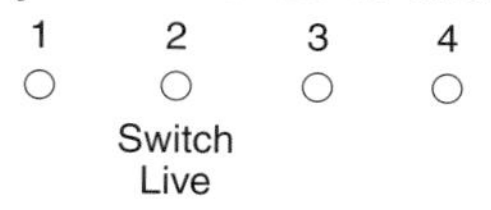

Remove Link 1–2

A 240V connection should be made to L and N on the Red Fyre control box.

Trianco TRO MK1

Floor mounted

Not suitable for use on sealed systems
Heat exchanger material, mild steel

Trianco TRO MK2 and 3

Floor mounted

Not suitable for use on sealed systems
Heat exchanger material, mild steel

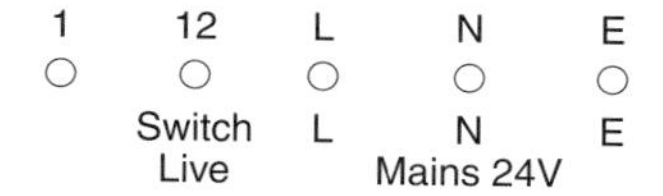

Remove Link 2–12

Trianco TRO 80 combination

Wall mounted

Suitable for use on sealed systems
Heat exchanger material, mild steel

Trianco TRO 110 combination

Floor mounted

Suitable for use on sealed systems
Heat exchanger material, mild steel

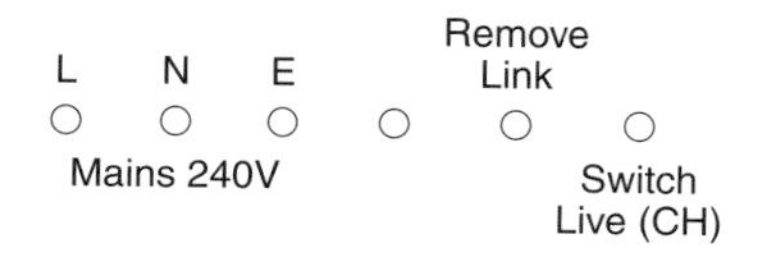

Trianco TRO BF range

Floor mounted

L N E SL

Perm Live N E Switch Live

Remove Link L–2 in 6 way plug-in terminal block

Not suitable for use on sealed systems
Heat exchanger material, mild steel

Trianco TSB system boiler

Wall mounted

L N E SL

240 V Mains Switch Live

Remove Link L–1 in 6 way plug-in terminal block

Suitable for sealed systems
Heat exchanger material, mild steel

Trianco Eurostar 2000

Floor standing

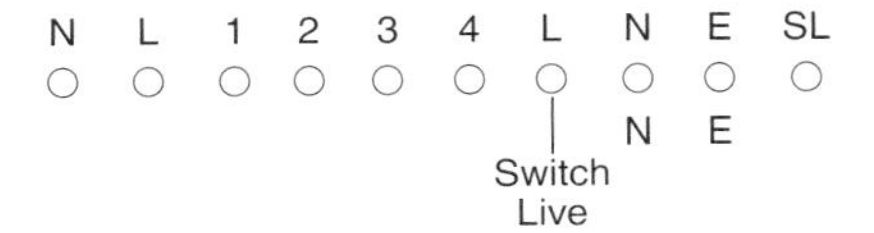

Fully pumped system only
Suitable for sealed systems
Heat exchanger material, mild steel
SEDBUK rating C

Trianco Eurostar Combi combination

Floor mounted

Suitable for use on sealed systems
Heat exchanger material, mild steel

External controls such as time clock with voltage free contacts or room thermostat should be connected between 3 and 6 after removing link.

Trianco Eurostar Combi

Floor standing

Fully pumped system only
Suitable for sealed systems
Heat exchanger material, mild steel
SEDBUK rating C

When fitting a room thermostat connect across T & H after removing link.

Trianco Eurostar Eco Combi

Floor standing

Remove link T–H when wiring room stat.

Fully pumped system only
Suitable for sealed systems
Heat exchanger material, mild steel
SEDBUK rating C

Trianco Eurostar FS 50/90, 95/115 external

Floor standing

Suitable for sealed systems
Heat exchanger material, mild steel
Integral frost thermostat fitted
SEDBUK rating C

Remove link L–SL if fitting external controls. A permanent live is required for frost protection.

Trianco Eurostar Standard

Floor mounted

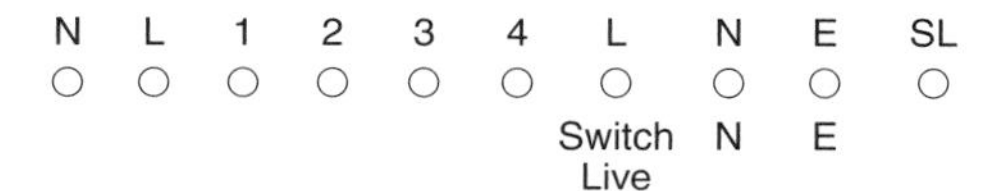

Suitable for use on sealed systems
Heat exchanger material, mild steel

Trianco Eurostar system

Floor mounted

Fully pumped systems only
Suitable for use on sealed systems
Heat exchanger material, mild steel

Trianco Eurostar Utility

Floor mounted

Suitable for use on sealed systems
Heat exchanger material, mild steel

Trianco Eurostar WM 50/65 external

Wall Mounted

Fully pumped system only
Suitable for sealed systems
Heat exchanger material, mild steel
Integral frost thermostat fitted
SEDBUK rating C

Trianco Eurostar WM 50/7

Wall mounted

The switched live or the live if no external controls are fitted should be connected to L.

Fully pumped system only
Suitable for sealed systems
Heat exchanger material, mild steel
SEDBUK rating C

Trianco Eurostar WM system boiler

Wall mounted

L N E

Switch Live N E

Fully pumped systems only
Suitable for use on sealed systems
Heat exchanger material, mild steel

Trianco Eurotrader

Floor standing

L N E

Switch Live N E Internal Connection

Suitable for sealed systems (with kit)
Heat exchanger material, mild steel
SEDBUK rating C

Trianco Utility

Floor standing

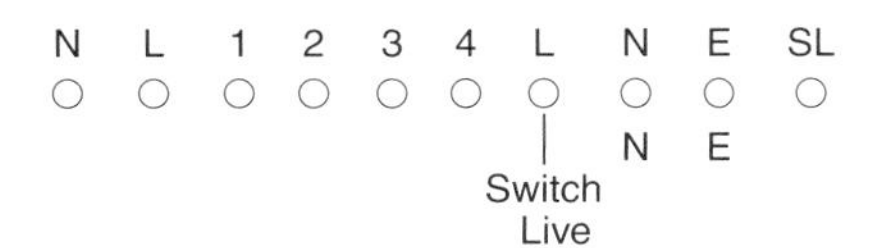

Fully pumped system only
Suitable for sealed systems
Heat exchanger material, mild steel
SEDBUK rating C

Trianco Utility Sealed system

Floor standing

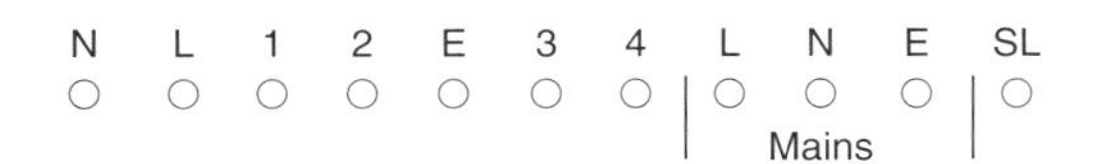

Fully pumped system only
Suitable for sealed systems
Heat exchanger material, mild steel
SEDBUK rating C

Warmflow Bluebird

Floor standing

Main connection via 3-way plug. Switched live to terminal 6. Terminals 7–8 remain linked.

Suitable for sealed systems
Heat exchanger material, mild steel
SEDBUK rating C

Warmflow Kabin Pak External

As Warmflow Bluebird

Warmflow Whitebird

As Warmflow Bluebird

Worcester Danesmoor 12/14, 15/19, 20/25, 26/32, 35/50, 50/70

Floor standing

Connect a mains supply to terminals L, N, E.
If fitting an external programmer and/or external controls refer to installation manual.

Suitable for sealed systems
Heat exchanger material, mild steel
SEDBUK rating C

Worcester Danesmoor DF

Floor mounted

Remove links L–L2 and L3–L4

Suitable for use on sealed systems using optional kit
Heat exchanger material, mild steel

Worcester Danesmoor PJ MK1 combination

Floor mounted

L N E 1 2 3

Mains 240V

Switch Live

Remove link L–3

Suitable for use on sealed systems using optional kit
Heat exchanger material, mild steel

Worcester Danesmoor PJ MK2

Floor mounted

With external controls

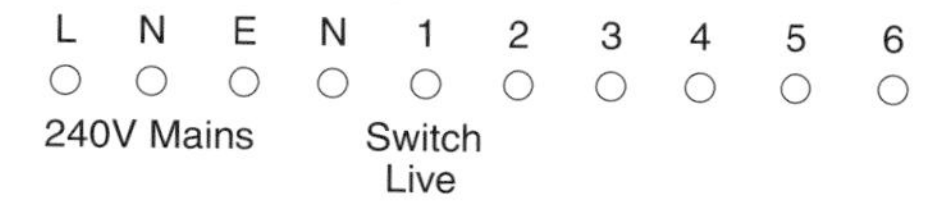

Remove link plug from programmer terminal strip

Suitable for use on sealed systems using optional kit
Heat exchanger material, mild steel

Worcester Danesmoor SLPJ

Floor mounted

With external controls

L	N	E	N	1	2	3	4	5	6
○	○	○	○	○	○	○	○	○	○
Mains lead supplied				Switch Live					

Remove link plug from programmer terminal strip

Suitable for use on sealed systems using optional kit
Heat exchanger material, mild steel

Worcester Danesmoor system

Floor standing

Connect a mains supply to terminals L, N, E.
If fitting an external programmer and/or external controls refer to installation manual.

Suitable for sealed systems
Heat exchanger material, mild steel
SEDBUK rating C

Worcester Danesmoor Utility

Floor standing

	○	1
	○	2
N	○	N
Switch Live	○	L
E	○	E
	○	3

Fully pumped system only
Suitable for sealed systems
Heat exchanger material, mild steel
SEDBUK rating C

Worcester Danesmoor WM12/19

Floor standing

Connect a mains supply to terminals L, N, E.
If fitting an external programmer and/or controls refer to installation manual.

Fully pumped system only
Suitable for sealed systems
Heat exchanger material, mild steel
SEDBUK rating C

Worcester Greenstar HE 12/22 condensing

Floor standing

L	E	N	1	2
○	○	○	○	○
L	E Mains	N	Room Stat or Timer	

Fully pumped systems only
Suitable for sealed systems
Heat exchanger material, stainless steel
SEDBUK rating A

External controls such as a time clock with voltage-free contacts or room thermostat should be connected across terminals 1–2 after removing link.

Worcester Heatslave combination 012–14 015–19 020–25

Floor mounted

Mains lead supplied. External controls, such as a time clock with voltage free contacts or room thermostat should be connected across terminals 2–4 after removing link. Room thermostat neutral can be connected to terminal N2.

Not suitable for use on sealed systems
Heat exchanger material, mild steel

Worcester Heatslave 2+ combination G40, G50

Floor mounted

Mains lead supplied. External controls, such as a time clock with voltage free contacts or room thermostat should be connected across terminals 2–4 after removing link. Room thermostat neutral can be connected to terminal N2.

Not suitable for use on sealed systems
Heat exchanger material, mild steel

Worcester Heatslave 12/14, 15/19, 20/25

Floor standing

Terminal block X2

6 5 4 3 2 1
○ ○ ○ ○ ○ ○

Fully pumped system only
Suitable for sealed systems
Heat exchanger material, mild steel
SEDBUK rating 12/14-D, others C

Connect room thermostat between terminals 2 and 3 after removing link. Room thermostat neutral should be connected to terminal 6.

Worcester Heatslave 26/32 Combination

Floor standing

Terminal block X2

6 5 4 3 2 1
○ ○ ○ ○ ○ ○

Fully pumped system only
Suitable for sealed systems
Heat exchanger material, mild steel
SEDBUK rating C

Connect room thermostat between terminals 2 and 3 after removing link. Room thermostat neutral should be connected to terminal 6.

Yorkpark Oilstar combination

Wall mounted

Fully pumped systems only
Suitable for sealed systems only
Heat exchanger material, stainless steel

Yorkpark Yorkstar condensing

Floor standing

Fully pumped systems only
Suitable for sealed systems with optional kit
Heat exchanger material, NK
Integral frost protection
SEDBUK rating A

Control wiring connections for when connected directly into burner control box

Control box	Perm live	Control live	N	E	Link information
Danesmoore TSV	L	HW ON-4 CH ON-6	N	E	Remove Links A–1 and A–2
Danfoss BHO 1		6*	N	E	
Danfoss BHO 15		9+	7	E	
Danfoss 57 F		1	3	E	Link 2–3
Danfoss 57 H		6	1	E	
Danfoss 57L		5	3	E	
DS 220		2	N	E	Link N–L1
Elestra		2	1	E	
Honeywell Protectorelay		1	2	E	
Landis & Gyr LAB 1		1+	2	E	
Landis & Gyr LOA 21		1+	2	E	
Nu-way ZLO		6	1	E	
Nu-way ZL2D		1	4	E	
Petercem MA28		10	4	E	
Satchwell DG		1	2	E	
Satronic TF 701 B		9+	8	E	
Satronic TF 830 N		9+	8	E	
Selectos D42		6	1	E	
Selectos JSS 1		6+	1	E	
Selectos JSS 2#		8	2	E	
Stewart TSV		as Danesmoor TSV			
Teddington DAS	L	11	N	E	Remove Link L–11
Thermoflex MC50		4	2	E	
Trianco TSV		as Danesmoor TSV			

* via limit thermostat
+via control and limit thermostats
connections into motor burner starter
Note: the earth connection may be a nut and bolt stud or screw into casing

11

Ancillary controls

Contents

Danfoss BEM 4000 Boiler Energy Manager

The Boiler Energy Manager (BEM) is an electronic controller which can be added to almost any central heating system to eliminate unnecessary boiler cycling and improve boiler seasonal efficiency.

The BEM 4000 consists of an electronic controller, an outdoor temperature sensor and a strap-on flow temperature sensor. It can be used in conjunction with a room thermostat (utilized as a frost stat) and TRVs. Boiler performance is improved when the flow temperature is varied according to the outside temperature. The outside temperature sensor should be mounted on the coldest elevation of the building away from opening windows and the boiler flue. The flow sensor should be fitted within 6 inches of the boiler outlet on the flow pipe, or on low-water content boilers it can additionally be fitted between the boiler and the bypass. A cylinder thermostat should be fitted as normal. All diagrams show Danfoss programmer and thermostats, but others can be used. In all cases, if a room thermostat is to be utilized as a frost or low-limit thermostat, connect to terminals 3 and 11.

In Figures 11.2 and 11.3 two boilers are shown – one with pump overrun and one without. Select diagram as appropriate, remembering to insert Link 2–4 when using boiler without pump overrun.

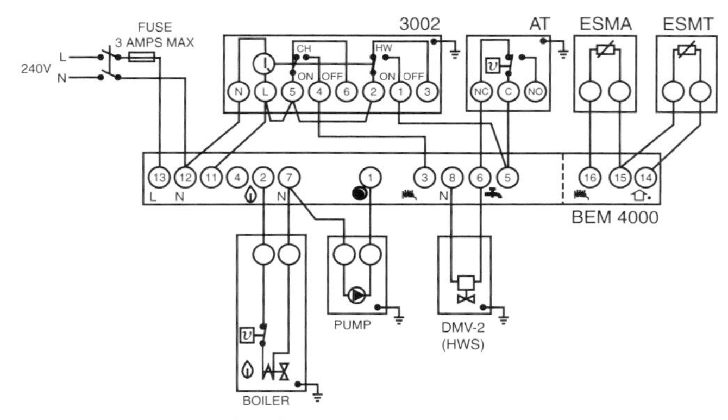

Figure 11.1 *Gravity hot water, pumped central heating*

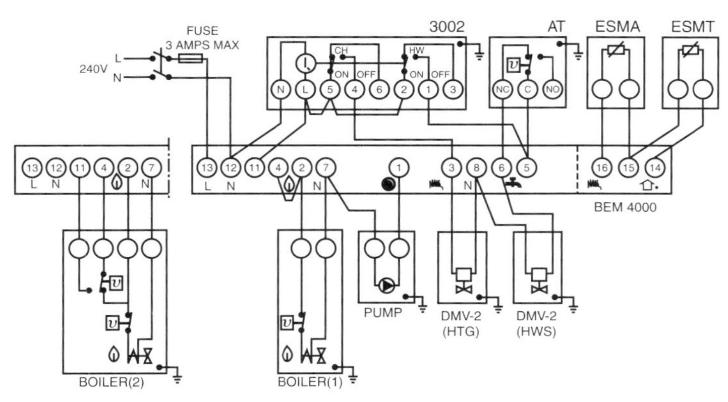

Figure 11.2 *Fully pumped system with two motorized zone valves*

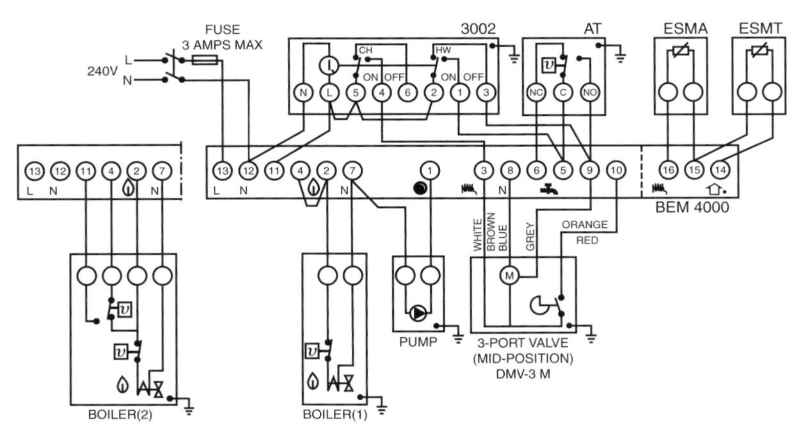

Figure 11.3 *Fully pumped system with mid-position valve*

Commissioning the BEM 4000

Hot water service

Set the programmer to hot water constant, central heating off, and cylinder thermostat to maximum. Set boiler to the design flow temperature of 82°C.

Lamps 1 and 2 will illuminate and boiler and pump will start, except in gravity hot water systems, in which the pump will not start. The hot water valve will open.

Central heating service

Set programmer to hot water off, central heating constant and boiler as above. Where a heating motorized valve is fitted, it will open. Illumination of lamp 3 signifies that the boiler has stopped and after a time lamp 1 will go off, indicating pump stopped.

After testing, set the programmer and cylinder thermostat to the required settings.

Fault finding

When the 'on' lamp on the front of the BEM 4000 does not illuminate, check the power supply. There should be 240V across terminals 12 and 13. If this is OK, check the built-in fuse on the rear of the BEM 4000, and if blown, there is a spare attached to the rear of the unit. If this fuse is OK the fault is probably within the electronics, and the BEM 4000 plug-in front plate should be replaced.

Where a zone valve will not open, check that 240V exists at the valve leads. If OK change valve, if not check programmer and other wiring.

In the event of difficulties, the BEM 4000 can be overridden by inserting a link between terminals 14 and 15, ensuring that the power is turned off first.

Danfoss Randall BEM 5000 Boiler Energy Manager

The BEM 5000 is an electronic control, which, when added to almost any central heating system, reduces unnecessary boiler cycling and improves boiler seasonal efficiency. It is compatible with most popular control systems including radiator thermostats, programmers, motorized valves and even room thermostats, although in systems having radiator thermostats it is recommended that the room thermostat be turned up to maximum or rewired as frost thermostats. Radiator thermostats are recommended, but are not mandatory. If they are fitted, then a by-pass valve should be fitted. If it is impractical to fit a by-pass valve, then one radiator should be left uncontrolled.

The BEM 5000 measures outdoor temperature and varies the temperature of water flowing to the radiators accordingly. This significantly improves boiler performance. A water temperature sensor is

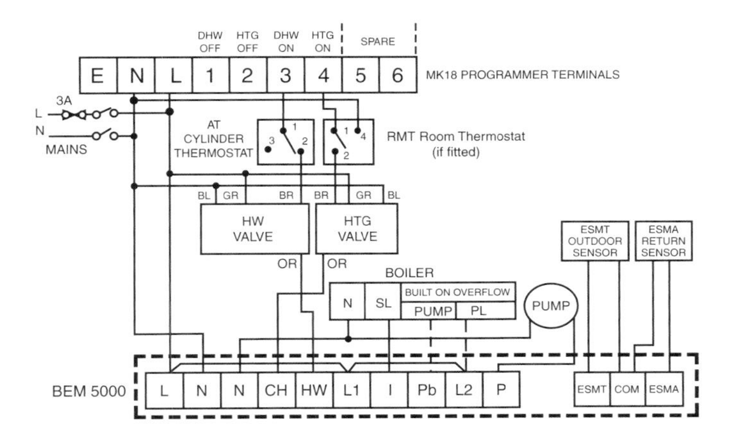

Figure 11.4

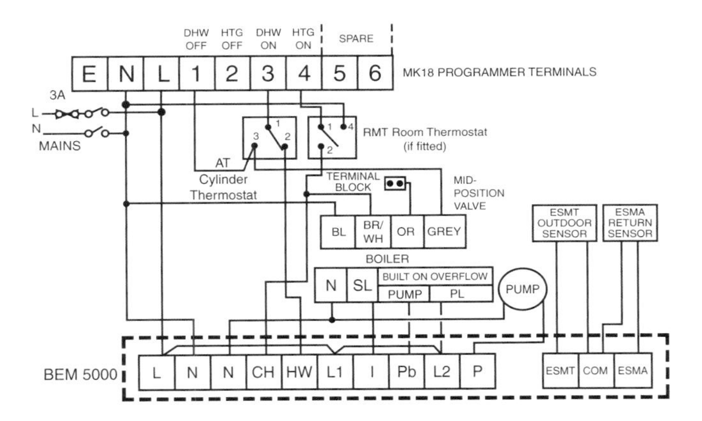

Figure 11.5

used to monitor temperature of water returning to the boiler. Any change in load on the system is measured and the boiler is controlled accordingly. To minimize energy loss from the boiler case and flue, the pump runs on after the boiler has stopped, circulating all useful heat to the system. When no more heat can be extracted, the pump is stopped. The BEM 5000 integrates the operation of heating and hot water. During periods of hot water demand, water temperature flowing to both water and heating circuits is boosted to boiler thermostat settings to ensure rapid recovery. Figures 11.4 and 11.5 show the BEM 5000 wired into a fully pumped system, one with two 2-port motorized valves and one with a 3-port, mid-position valve. It can also be utilized on a combination boiler.

Dataterm optimiser

Dataterm is a microprocessor-based energy management system and can be used as part of a new system or to replace traditional room thermostats and programmer. The fully automatic optimum start function takes into consideration the current weather and calculates the boiler start time eliminating fuel waste through fixed boiler start times. Dataterm is suitable for any conventional oil or gas wet central heating system complete with hot water control, as well as other applications.

The control pack consists of two units, a low-voltage programmable room thermostat and a controlling power pack located in a convenient position, e.g. adjacent to the boiler or motorized valves.

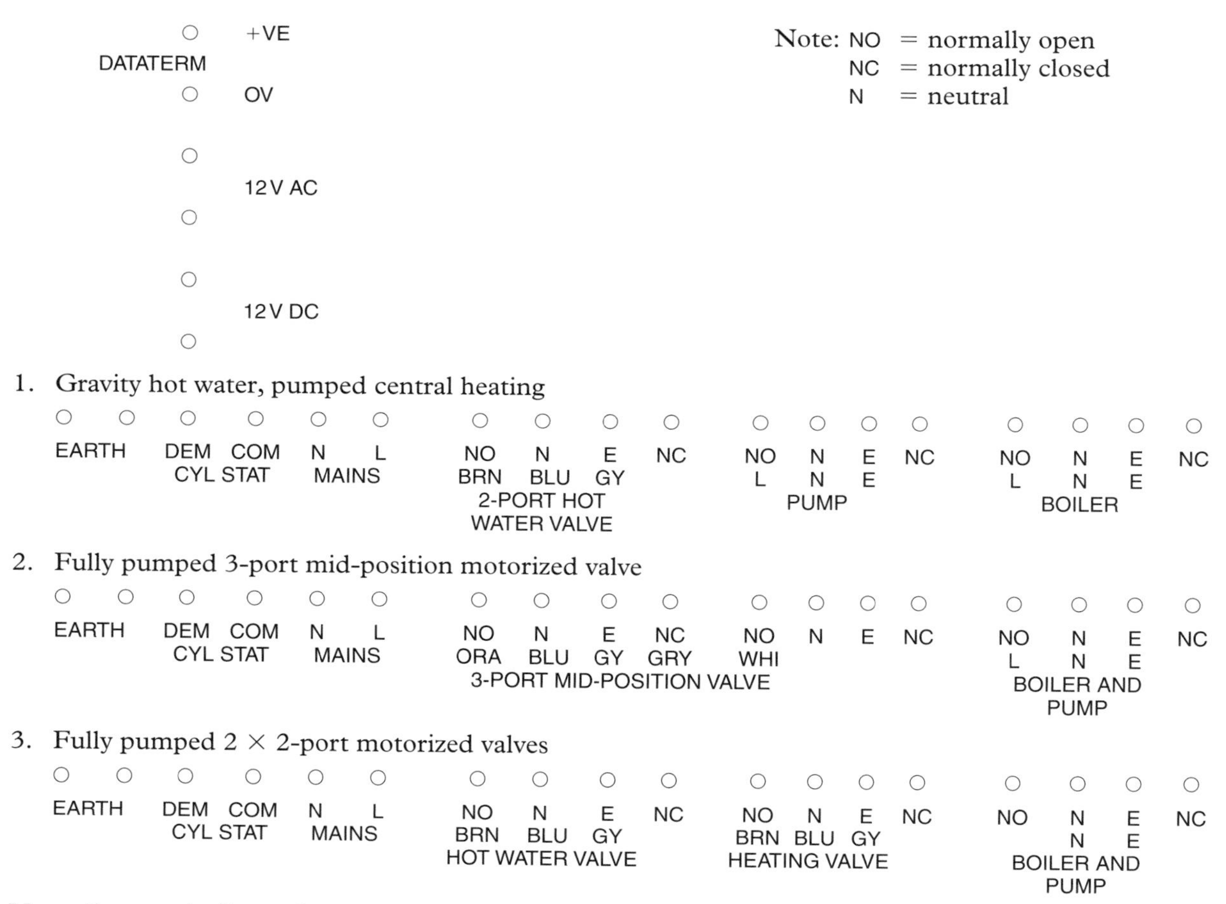

Note: Connect boiler and pump lives with motorized valve oranges in separate connector. Connect motorized valve greys into mains L.

Honeywell AQ 6000

The AQ 6000 is an outside temperature compensator system for use in domestic properties and can be used on new installations or to upgrade existing systems, making use of existing valves that may be installed. The AQ 6000 is available in three packs – an Upgrade pack, a Standard pack or a Modulating pack. The Modulating pack is for use on boilers of 90 000–150 000 BTU and provides control of the heating circuit by modulating a mixing valve. There are five systems – A, B, C, D and E. Systems D and E use the modulating mixing valve and are not shown.

The AQ 6000 consists of a control unit, a room unit and outside, water supply and domestic hot water sensors. All wiring is terminated in the control unit along with boiler, pump and any valve wiring.

A built-in start-up operating sequence allows the system wiring to be tested. Before turning on power to the system, remove the room unit from its mounting bracket. Turn on the power and the control unit will power the valve, pump and boiler.

When the system is operational actual temperatures can be displayed on the room unit. When the green enquiry button is pressed, 'T1' will be displayed at the left side of the display and the temperature in °C on the right. Repeated use of this button will display other system temperatures. The codes are listed in Table 11.1. To restore the display to its normal operation press the enquiry button repeatedly until the 'T' is no longer displayed and in its place is the current weekday number.

Table 11.1

Code	Temperature measured
T1	Room temperature
T2	Boiler/mixed water temperature
T3	Outside temperature
T4	System C – not used (will display 0°C) Systems A and B – domestic hot water temperature

When the room unit is connected and the supply switched on, an indication of any faults within the system can be displayed by pressing any four buttons on the unit. A fault code is displayed as an 'F' followed by a number. The different numbers correspond to the faults listed in Table 11.2. No 'F' on the display means that no fault has been observed by the controller. To clear the fault code, press any button.

Table 11.2

Code	Indicates	Possible causes	Corrective action
F1	Hot water has not reached 30°C within 40 minutes of start-up	Boiler not firing	Check wiring Check appliance function
		Pump not running	Check wiring
		Valve being driven closed	Reverse leads 1 and 2 on the actuator *or* on the boiler unit
		Valve in closed position	Check wiring Check actuator
F2	Boiler/mixed water temperature sensor fault	Faulty sensor wiring	Check wiring for open or short circuits
		Faulty sensor	Change sensor
F3	Outside temperature sensor fault	Faulty sensor wiring	Check wiring for open or short circuits
		Faulty sensor	Change sensor
F4	Failure of the communications link between the room and the boiler room	Room unit not securely seated in the mounting bracket	Check seating of room unit in its bracket
		Faulty wiring	Check for open or short circuits
		Wire connected the wrong way round	Reverse wiring connection at the boiler unit or room unit
		No power at boiler unit	Check boiler supply and 50 mA fuse on boiler unit board
F5	Limit or domestic hot water temperature sensor fault	Faulty wiring	Check wiring for open or short circuits On systems C and D, check that the resistor is still in place between terminals 3 and 4 of the boiler unit On other systems, check that the resistor has been removed
		Faulty sensor	Change sensor

System A: Boiler control with domestic hot water control using two motorized zone valves

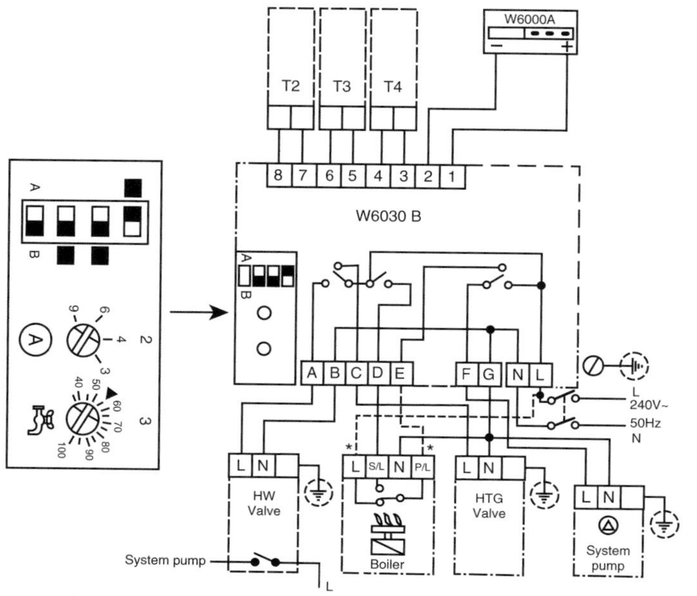

Figure 11.6

The following adjustments must be made **before** power is applied to the system:

1. System selection switches.
2. Burner cycles per hour (3 to 9), typical setting: 6.
3. Domestic hot water (35 up to 100°C), typical setting: 55°C.

Note: The switch A–B should be in position A for the domestic hot water to be off when the system is controlling at the Economy level. The switch should be in position B for the domestic hot water to be continuously serviced.

System B: Boiler control with domestic hot water control using a 3-port motorized valve

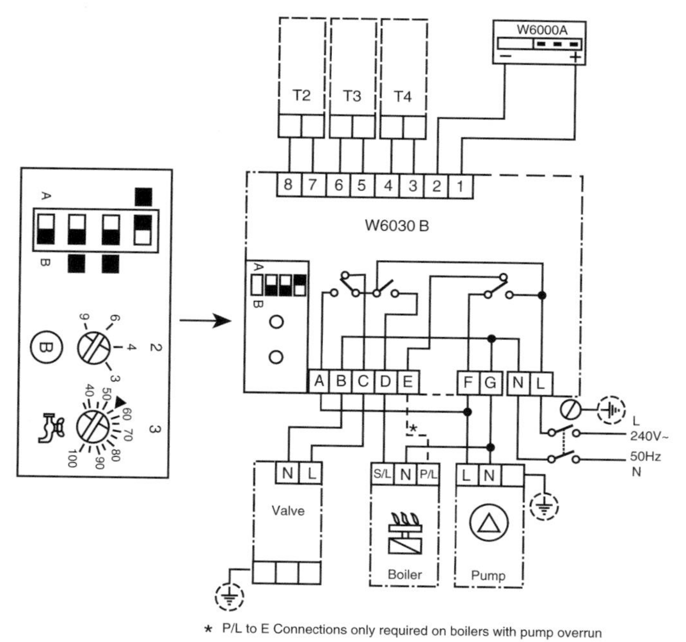

Figure 11.7

The following adjustments must be made **before** power is applied to the system:

1. System selection switches.
2. Burner cycles per hour (3 to 9), typical setting: 6.3.
3. Domestic hot water (35 to 100°C), typical setting: 55°C.

Note: The switch A–B should be in position A for the domestic hot water to be off when the system is controlling at the Economy level. The switch should be in position B for the domestic hot water to be continuously serviced. Wiring shows a priority valve. Where a mid-position valve is used the White (or Brown) and Grey go to terminal C.

System C: Boiler control without domestic hot water control

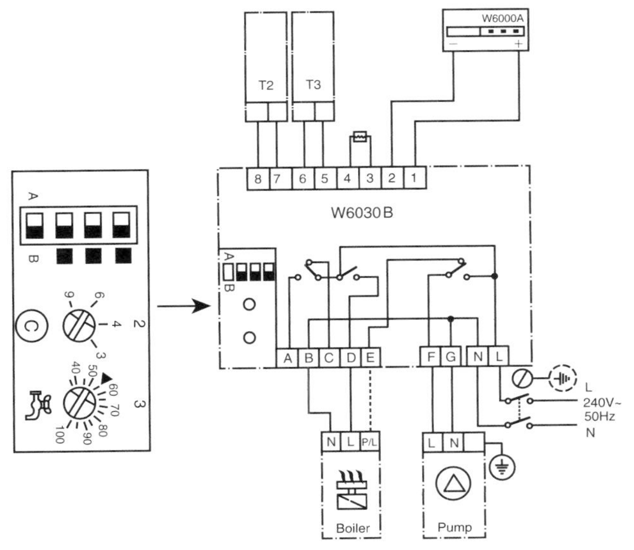

Figure 11.8

The following adjustments must be made **before** power is applied to the system:

1. System selection switches.
2. Burner cycles per hour (3 to 9), typical setting: 6.

Note: Only recommended for systems up to 70 000 BTU. Domestic hot water sensor, T4, is not used.

Honeywell Y604A Panel and Timed Sundial Plans

The Y604A Panel and Panel Timed Sundial Plans are fully assembled, pre-plumbed and pre-wired control sets and comprise of either a Honeywell S or Y Plan, or Timed S or Y Plan, plus pump and isolating valves in a multipoise carrier bracket.

Table 11.3 *External wiring connections – panel plans*

		S Plan terminal	Y Plan terminal
Room thermostat	Common	4	4
	Demand	5	5
	Neutral	N	N
Basic boiler	Live	7	8
	Neutral	N	N
	Earth	E	E
Boiler with pump overrun	Remove link 7–9 in junction box		
	Live	L	L
	Neutral	N	N
	Earth	E	E
	Boiler on	7	8
	Pump live	9	9
Programmer	Live	L	L
	Neutral	N	N
	Heating on	4	4
	Hot water on	6	6
	Hot water off	–	7
Frost thermostat	Common	1	1
	Demand	5	8

Table 11.4 *External wiring connections – panel timed plans*

		S Plan terminals	Y Plan terminals
Room thermostat	Common	4	4
	Demand	5	5
	Neutral	N	N
Basic boiler	Live	7	8
	Neutral	N	N
	Earth	E	R
Boiler with pump overrun	Remove link 7–9 in junction box		
	Live	L	L
	Neutral	N	N
	Earth	E	E
	Boiler on	7	8
	Pump live	9	9
Frost thermostat	Common	1	1
	Demand	5	8

Honeywell Y605B Panel Link-Fuel Timed Sundial Plan

The Y605B Panel Link-Fuel Timed Sundial Plan is an assembled, pre-plumbed and pre-wired control set. Designed for use with fully pumped wet central heating systems linking together a solid-fuel boiler with either a gas, oil or electric boiler. Full time and temperature control of the system is achieved by use of the programmer, room and cylinder thermostats.

The system selector box offers the choice of heating by means of solid-fuel only, where the link-fuel boiler will not be used, or as a link-up system where both the solid-fuel and link-fuel boilers will be used.

A second selection allows choice between dissipating excess heat either into the heating or the domestic hot water circuits. It is recommended that heating be selected during the winter and hot water in spring or autumn, whenever the solid-fuel appliance is in use.

When commissioning, set the required on/off times of the system on the programmer, set required temperature for domestic hot water, and the room thermostat to the required room temperature. The triple aquastat limit thermostat is factory set but may need to be adjusted.

Control operation

A. When solid fuel only has been selected. The link-fuel valve will be closed and the solid-fuel valve will be open and the system will operate as follows:

1. No demand within the system slumber valve open to allow full bore gravity circulation to the slumber circuit. Safe operation of solid-fuel appliance without overheating of space or domestic hot water. Heating valve and domestic hot water valve closed, pump off.
2. Heating only: pump on, slumber and hot water valves closed, heating valve open to allow heating to be provided. Water pumped around the slumber circuit through adjustable gate valve.
3. Hot water only: pump on, slumber and heating valves closed, hot water valve open to allow circulation to the domestic hot water. Water pumped around the slumber circuit through adjustable gate valve.
4. Both heating and hot water: pump on, slumber valve closed, heating and hot water valves open to allow circulation throughout the whole of the system. Water pumped around the slumber circuit through the adjustable gate valve.

5. If the temperature of the water exceeds the high-limit setting at any time the safety controls will take effect. Then pump on, slumber valve closed whilst either heating or hot water valve will open to allow for excess heat to be dissipated into either the hot water or heating and slumber circuits, as selected. The high-limit thermostat will automatically reset when the temperature of the water has reduced and the system will return to normal operation.

B. When the link-fuel has been selected the system operates in conjunction with the limit thermostats:

1. When flow temperature from solid-fuel appliance is below the mid-limit, the link-fuel boiler is brought into use. The slumber valve is closed, link-fuel and solid-fuel valves open and either heating, or hot water, or both, valves open to satisfy demand of the system. Water is supplied to the slumber circuit through the adjustable gate valve and both the pump and link-fuel boiler are on. Operation of the system is switched between the solid-fuel and link-fuel by the mid-limit thermostat.
2. When flow temperature from solid-fuel appliance is between the mid- and high-limit temperatures then the link-fuel boiler is switched off. Full use is made of the output from the solid-fuel appliance only, hence achieving maximum use of the solid-fuel appliance and economizing on other fuels. Link-fuel valve closed.
3. If the flow temperature from the solid-fuel appliance exceeds the high-limit setting then the link-fuel boiler is switched off. Slumber and link-fuel valves close, solid-fuel valve opens as well as either heating or hot water valve, depending upon selection of either heating or hot water for dissipation for excess heat. Pump is on to remove excess heat through the chosen circuit and also through the slumber circuit via the adjustable gate valve.
4. The low-limit thermostat provides control of the solid-fuel appliance at low temperature.
   (a) Temperatures below the low-limit setting, solid-fuel valve will be closed.
   (b) Temperatures above the low-limit setting, solid-fuel valve will be open.
   (c) Any requirement for heating or hot water will be met by the link-fuel boiler exactly as Item 1 above.

Note: When pump is off, slumber valve is open
When pump is on, slumber valve is closed.

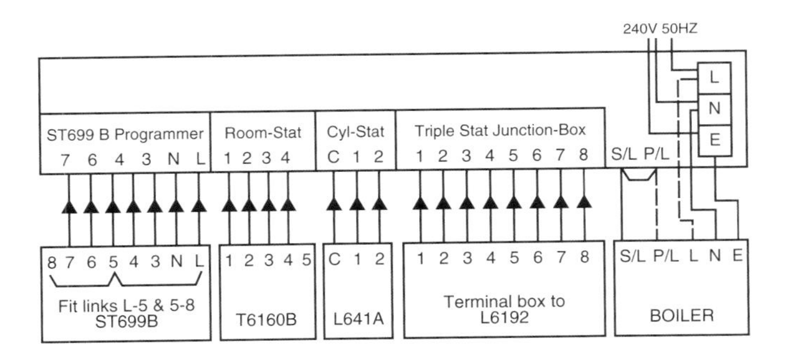

Figure 11.9

When using pump overrun boiler do not fit link S/L to P/L but connect L to L and P/L to P/L.

Randall EBM 2.1 Boiler Efficiency Control

The EBM 2.1 Boiler Efficiency Control is designed, with one simple user adjustment, to maximize the efficiency of gas fired wet domestic central heating systems. This is for systems which incorporate time and temperature controls (TRV or room thermostat and cylinder thermostat) and motorized valves. Suitable for both new and existing installations incorporating conventional or condensing boilers in either gravity or fully pumped systems.

The following pages show various systems and their wiring diagrams, however, before turning to these it is important to note the following:

1. The EBM 2.1 should be mounted where it is clearly accessible.
2. The OTS outside sensor should be mounted on an outside wall on the coldest side of the house (usually north facing) about 18 inches below eaves height and away from chimneys, flues or windows, etc. Connection must be made with cable suitable for 240 V as follows:
   - Terminal 1 on the OTS to terminal 11 on the EBM wallplate.
   - Terminal 2 on the OTS to terminal 10 on the EBM wallplate.
3. The BRS boiler return sensor should be mounted on the common return to the boiler between the boiler and by-pass. In gravity hot water systems it should be mounted on the heating return. Connect to EBM as follows:
   - Brown lead to terminal 12 on the EBM wallplate.
   - Blue lead to terminal 10 on the EBM wallplate.
4. The cylinder thermostat and any motorized valves shown are not included in the kit.
5. The pump should be connected to terminal 5 of the EBM in all cases, even if the boiler has its own pump connection or pump overrun facility.
6. In gravity hot water systems, it is necessary to cut an internal link in the EBM by prizing off the back cover and cutting the grey link wire located inside the housing on the left.
7. The boiler thermostat should be set to maximum.

Commissioning the EBM 2.1

1. Set the response rate setting on the EBM 2.1 to 10, **except for condensing boilers,** where the response rate should be set at 5.
2. Set the boiler thermostat to maximum.
3. Set the cylinder thermostat to maximum.
4. Set the room thermostat (if fitted) to maximum.
5. Turn on the power to the system. The 'Saving' light will illuminate.
6. Set the programmer on HW 'ON'. The 'HW' and 'Pump' lights will illuminate. The boiler will fire and the pump will run. The 'Saving' light will switch off.
7. Allow the system to warm for about 15 minutes and turn the cylinder thermostat to minimum. The 'HW' light will switch off and the boiler will stop. The pump (and 'Pump' light) may continue to run depending on the return temperature (see note above for gravity primary systems). The 'Saving' light will switch on.
8. Set the programmer to CH 'ON'. The 'CH' light will illuminate. The 'Saving' light will switch off and the boiler and pump will run.

   Note: If the outside temperature is greater than approximately 17°C, it will not be possible to run the CH service as outlined above as it will not switch on. If this is the case, disconnect the sensor lead from terminal 11 on the wallplate which will force the CH service on. After checking, the sensor lead should be reconnected to terminal 11.
9. When the system is up to temperature, the boiler will stop and the 'Saving' light will illuminate. The pump will continue to run. When the system temperature drops sufficiently the boiler will fire again and the 'Saving' light will switch off.

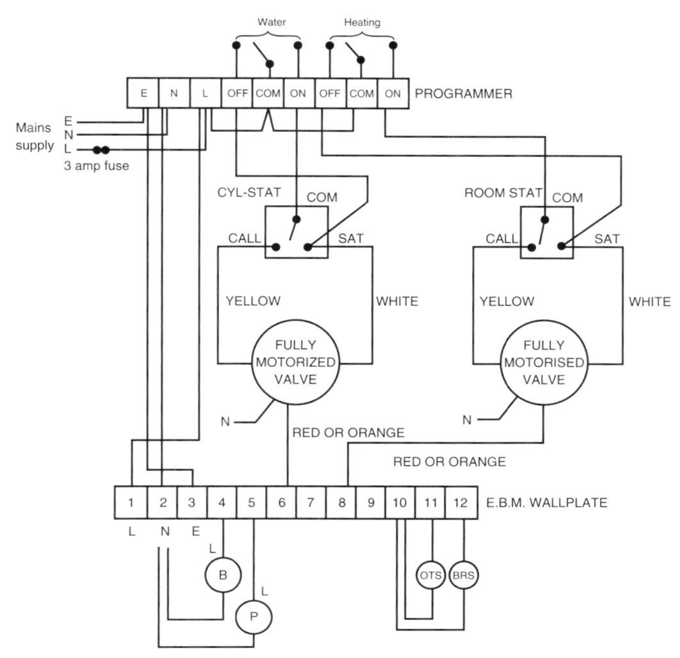

Figure 11.10 *Fully pumped system with two motor open, motor closed motorized valves*

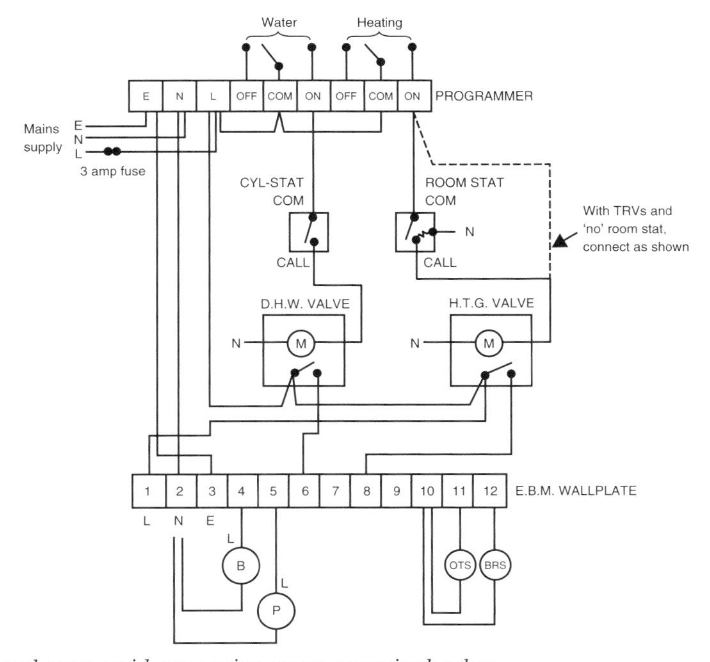

Figure 11.11 *Fully pumped system with two spring-return motorized valves*

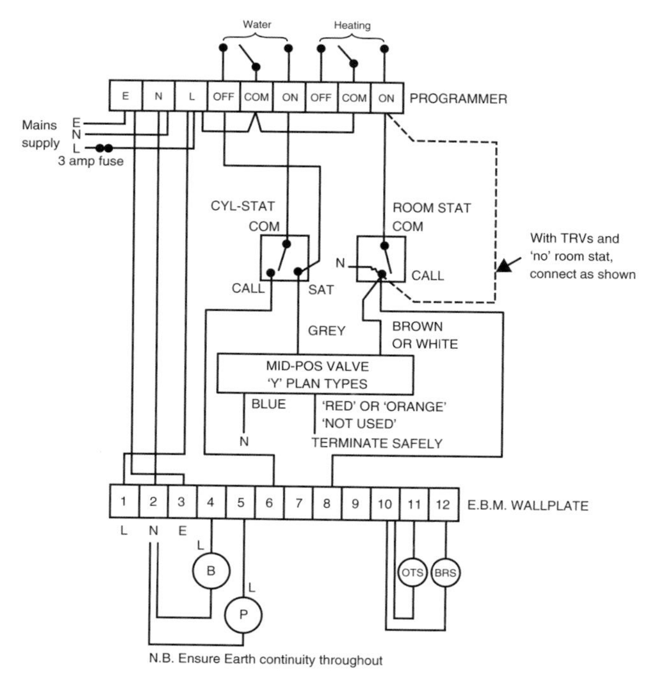

Figure 11.12 *Fully pumped system with mid-position motorized valve*

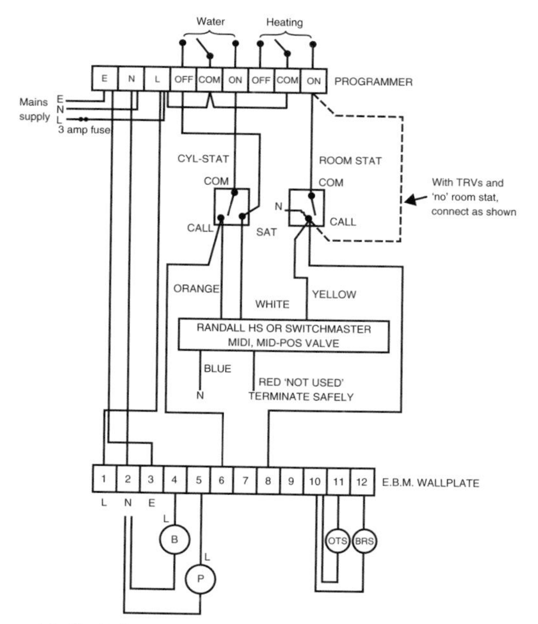

Figure 11.13 *Fully pumped system with Switchmaster 'Midi' or Randall HS mid-position valve*

Sunvic Clockbox and Clockbox 2

The Clockbox is a complete plug-in central heating control pack with programming facility into which the motorized valve and cylinder stat are plugged in to. The boiler, pump, room stat and mains are wired in to their respective terminals.

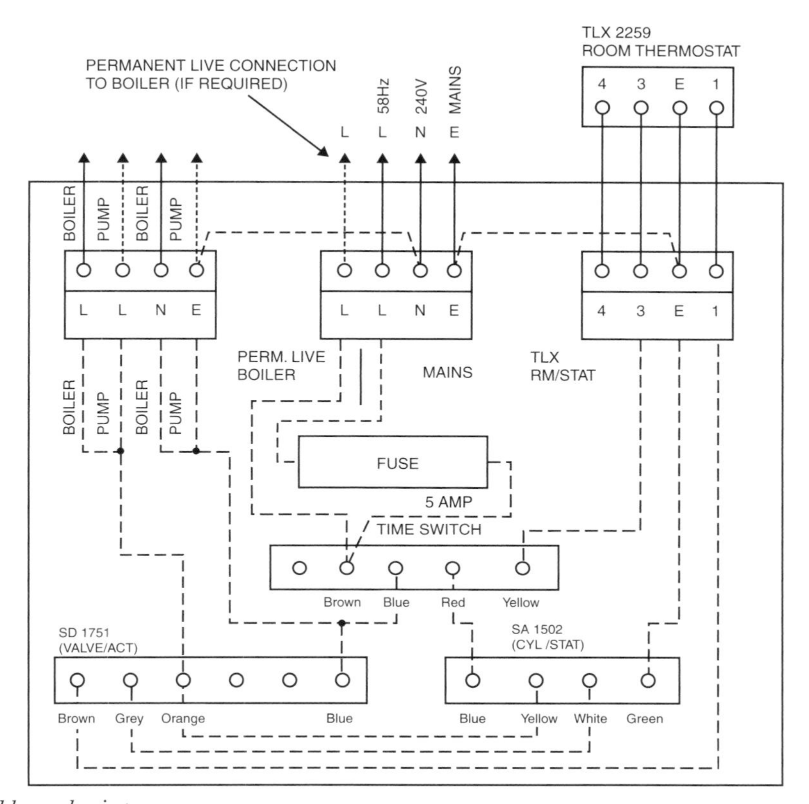

Figure 11.14 *Clockbox – basic programmer*

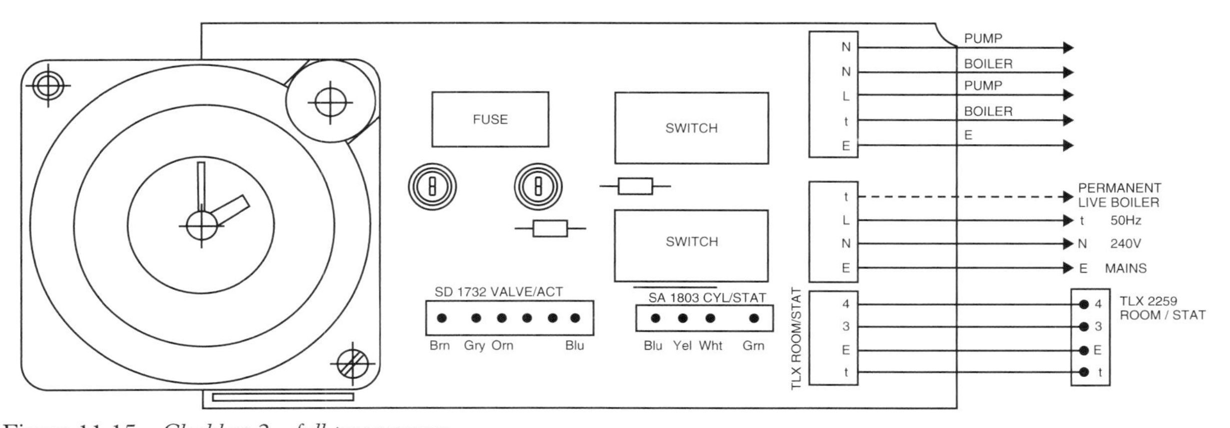

Figure 11.15 *Clockbox 2 – full programmer*

12

Wiring system diagrams

Contents

Gravity hot water, pumped central heating

Fully pumped systems

Figure 12.23 With a mid-position 3-port valve with standard colour flex conductors. Provides full temperature and programming control

Figure 12.24 With a 2-port motor open, motor close motorized valve to both central heating and hot water circuits. Provides full temperature and programming control

Figure 12.25 With a 2-port motor open, motor close motorized valve and a 2-port spring return motorized valve to control central heating and hot water circuits

Figure 12.26 With a 2-port spring return motorized valve to both central heating and hot water circuit when no permanent live is available for valve auxiliary switches. Provides full temperature control

Figure 12.27 With a 2-port spring return *normally open* motorized valve to both central heating and hot water circuits

Figure 12.28 Showing the addition of a 2-port spring return motorized valve to an existing system for zone control

Figure 12.29 Showing multiples of 2-port spring return motorized valves wired to one programmer, as may be used in a large property for zone control

Figure 12.30 Showing multiples of 2-port motor open, motor close motorized valves wired to one programmer, as may be used in a large property for zone control

Figure 12.31 With one boiler and zone control using one pump for each zone. No motorized valves

Figure 12.32 With two or more pump overrun boilers and one pump using relays as required

Figure 12.33 With pump to each circuit and no motorized valves

Figure 12.34 With pump overrun boiler, pump to each circuit, room and cylinder thermostats, programmer and relay. No motorized valves

Figure 12.35 With pump overrun boiler, pump and 2-port spring return motorized valves to each circuit, room and cylinder thermostats, programmer and relay

Figure 12.36 ACL Biflo System MK1 with time clock

Figure 12.37 ACL Biflo System MK1 using a programmer with voltage free contacts. Provides full temperature and programming control

Figure 12.38 Danfoss 2.2 and 2C System

Figure 12.39 Drayton Plan 1 with Drayton TA/M2 actuator

Figure 12.40 Drayton Flowshare 5 System with TA/M4 actuator, RB1 relay box and programmer

Figure 12.41 Drayton Flowshare 5 System with Drayton TA/M4 actuator, RB2 relay box and programmer

Figure 12.42 Drayton Plan 7 System with two TA/M2A actuators providing full temperature and programming control

Figure 12.43 Homewarm Manual System

Figure 12.44 Homewarm Auto System with Switchmaster VM5 actuator

Figure 12.45 Honeywell 'Y' Plan System using the V4073 6-wire, 3-port motorized valve with integral relay giving full temperature control

Figure 12.46 Honeywell 'Y' Plan System using the V4073 6-wire, 3-port motorized valve with integral relay giving full temperature and programming control

Figure 12.47 Landis & Gyr LGM System

Figure 12.48 SMC Control Pack 2 System with one boiler and a pump to both central heating and hot water circuits. No motorized valves. Provides full temperature and programming control

Figure 12.49 SMC Control Pack 2 System with pump overrun boiler and a pump to both central heating and hot water circuits. No motorized valves. Provides full temperature and programming control

Figure 12.50 Sunvic Duoflow System with RJ 1801 relay box and programmer

Figure 12.51 Sunvic Duoflow System with RJ 2801 relay box and programmer

Figure 12.52 Sunvic Duoflow System with RJ 2802 or RJ 2852 relay box and time clock

Figure 12.53 Sunvic Duoflow System with RJ 2802 or RJ 2852 relay box and programmer

Figure 12.54 Switchmaster Midi System

Various frost protection thermostat wiring

Supplementary wiring diagrams

Gravity hot water, pumped central heating

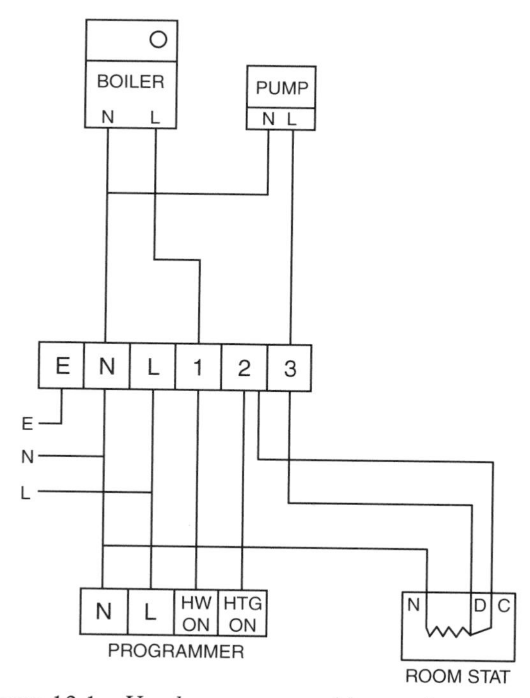

Notes

a) Known as Drayton Plan 2.
b) Suitable only for Basic programming.
c) Connect frost thermostat across junction box terminals L–1 for hot water or double-pole frost thermostat L–1 and L–3.

Figure 12.1 *Usual arrangement with room thermostat controlling pump*

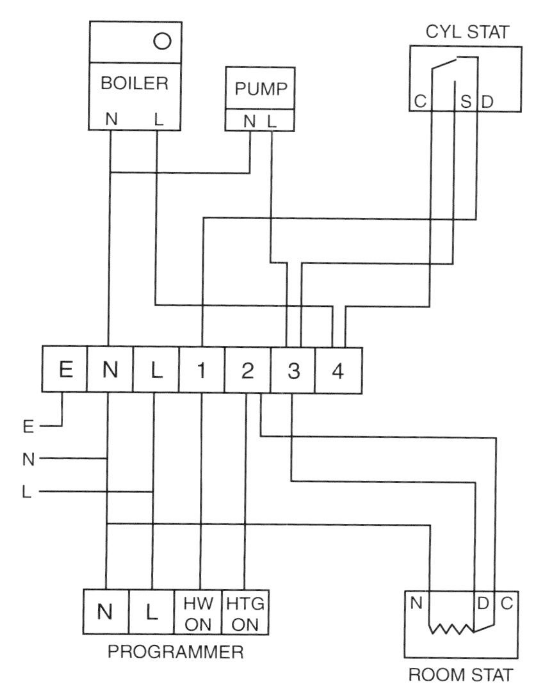

Figure 12.2 *With a cylinder thermostat to control hot water temperature*

Notes

a) Known as Honeywell 'A' Plan.
b) Suitable only for Basic programming.
c) Cylinder thermostat will only control hot water temperature when 'Hot Water Only' is selected on programmer.
d) Connect frost thermostat across junction box terminals L–4 for hot water or use double-pole frost thermostat L–4 and L–3.

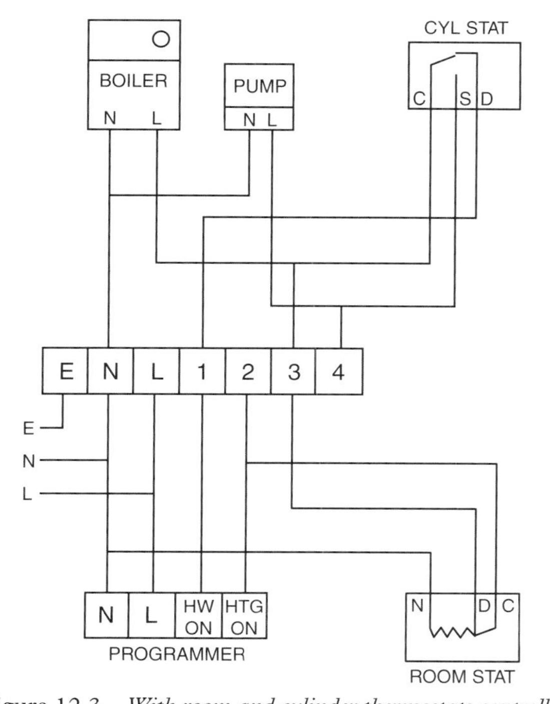

Figure 12.3 *With room and cylinder thermostats controlling boiler and pump giving hot water priority*

Notes

a) Suitable for Basic or Full control programming but no temperature control of hot water when heating is on.
b) Heating will not work until cylinder thermostat is satisfied.
c) Connect frost thermostat across junction box terminals L–1.

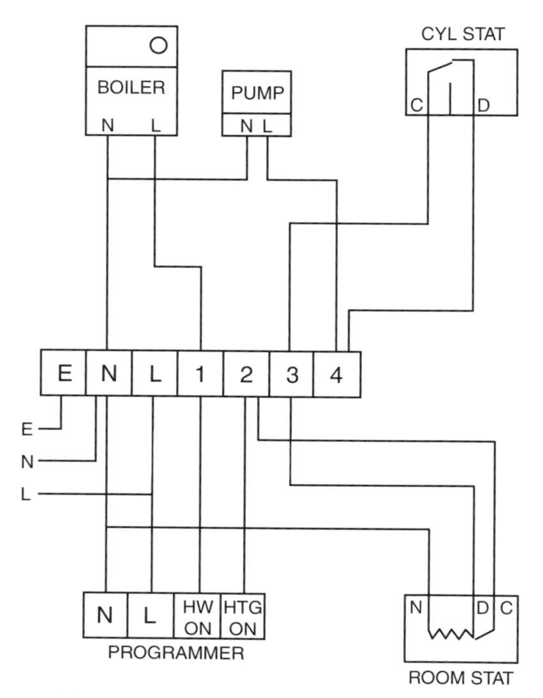

Notes

a) This method provides for the heating of the hot water to the cylinder thermostat temperature before the heating pump can function. This is ideal for systems where the heating pump will 'starve' the cylinder.
b) No hot water temperature control. Cylinder thermostat serves only to delay heating until cylinder is hot.
c) Suitable for Basic programmers only.
d) Connect frost thermostat L–1 for hot water or double-pole frost thermostat L–1 and L–4.

Figure 12.4 *With room and cylinder thermostats both controlling pump, giving hot water priority*

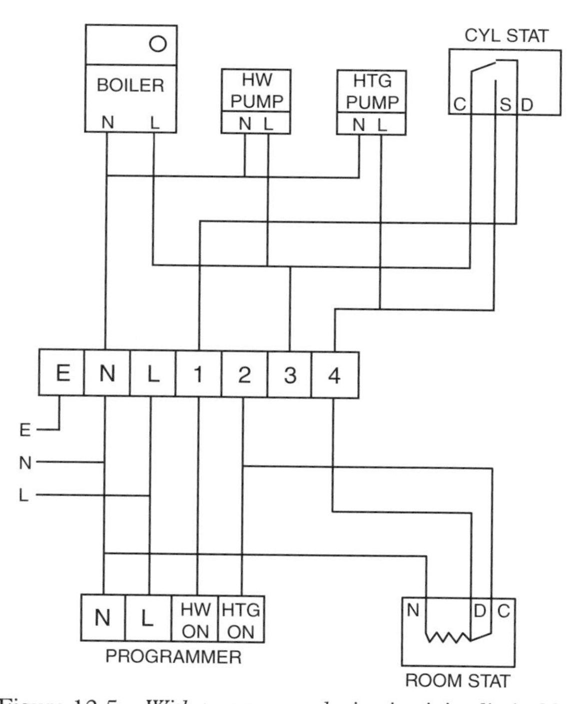

Notes

a) Suitable only for Basic programming.
b) Hot water temperature control only possible when 'Hot Water Only' selected.
c) Connect frost thermostat across junction box terminals L–3 for hot water only or double-pole frost thermostat L–3 and L–4.

Figure 12.5 *With pump on each circuit, giving limited hot water temperature control*

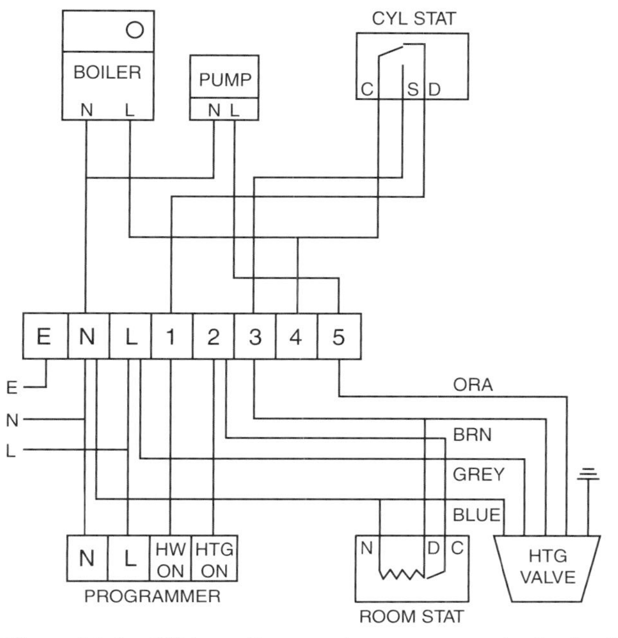

Notes

a) Known as Honeywell System L.
b) Suitable only for Basic programming.
c) Hot water temperature control only possible when 'Hot Water Only' selected.
d) Connect frost thermostat across junction box terminals L–3.

Figure 12.6 *With one 2-port spring return motorized valve in central heating circuit, giving limited temperature control of hot water*

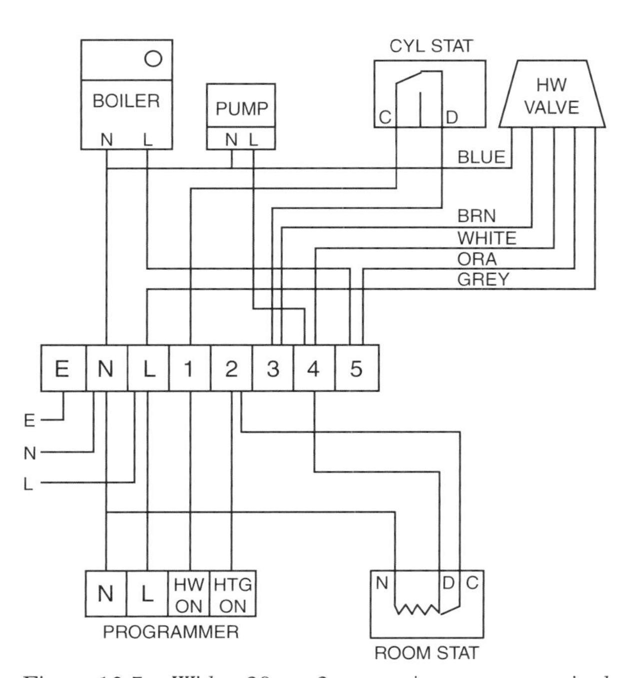

Notes

a) Known as Honeywell 'C' Plan.
b) Suitable for Basic or Full programming.
c) Connection of grey can be made to junction box terminal 1 (Hot Water On) if no live at junction box.
d) Motorized valve must have changeover auxiliary switch.

Figure 12.7 *With a 28mm 2-port spring return motorized valve to hot water circuit, giving full temperature and programming control*

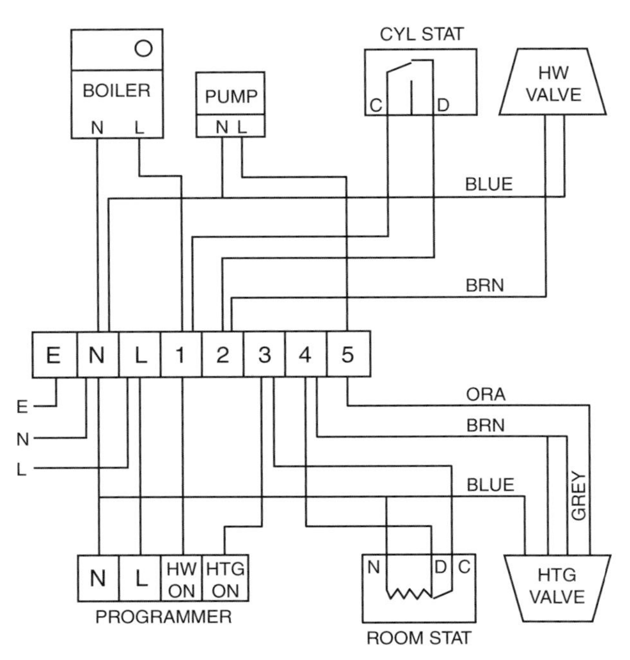

Figure 12.8 *With a 2-port spring return motorized valve to both central heating and hot water circuits, giving full temperature control*

Notes

a) Suitable only for Basic programming.
b) Connect frost thermostat across junction box terminals L–1 for hot water only or double-pole frost thermostat L–1 and L–4.

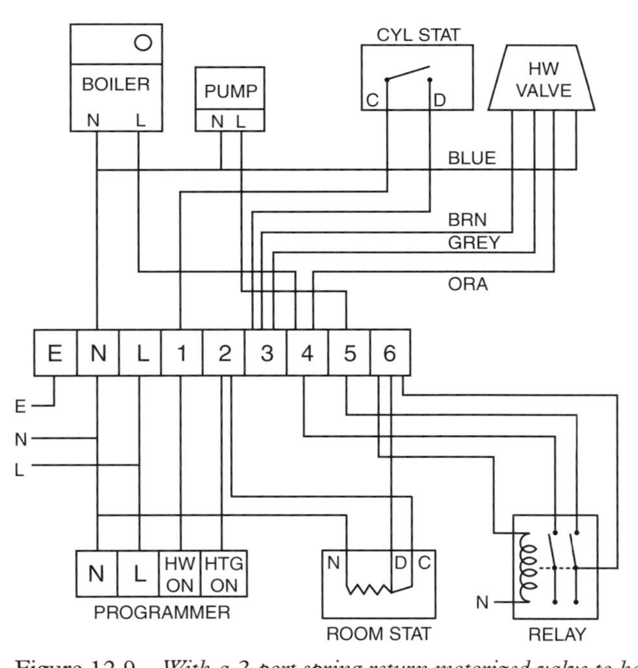

Figure 12.9 *With a 2-port spring return motorized valve to hot water circuit and relay, giving full temperature and programming control*

Notes

a) Suitable for Basic or Full programming.
b) Requires a DPDT relay.
c) Connect frost thermostat across junction box terminals L–6.

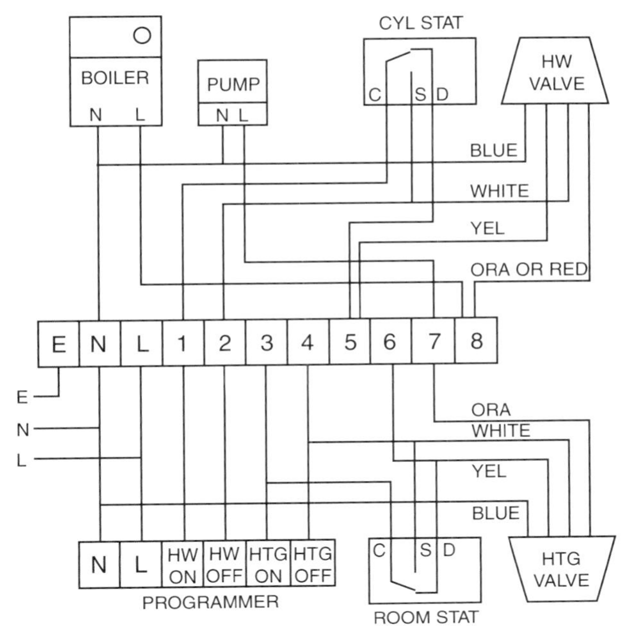

Figure 12.10 *With a 2-port motor open, motor close motorized valve to both central heating and hot water circuits, giving full temperature control*

Notes

a) Motorized valve on heating to stop gravity circulation.
b) System requires changeover room and cylinder thermostats.
c) Suitable for Basic control only but programmer must have 'Offs' to heating and hot water.
d) For frost protection see special diagram, page 258

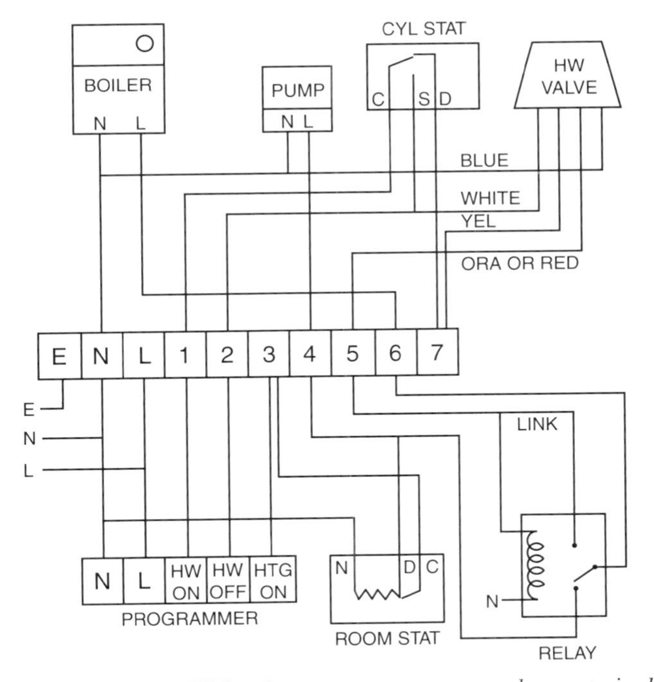

Figure 12.11 *With a 2-port motor open, motor close motorized valve to hot water circuit and relay, giving full temperature and programming control*

Notes

a) Suitable for Basic or Full programming but 'Hot Water Off' signal required regardless of whichever is used.
b) Requires an SPDT relay.
c) Connect frost stat across junction box terminals L–4.

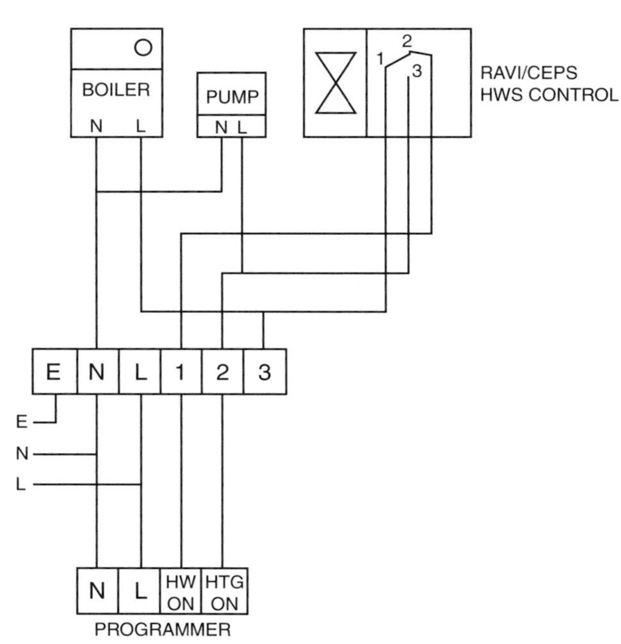

Figure 12.12 *Danfoss Plan 1.1 and Plan 2A*

Notes

a) Suitable only for Basic programming.
b) Connect frost thermostat across junction box terminals L–3 for hot water or double-pole frost thermostat L–3 and L–2.

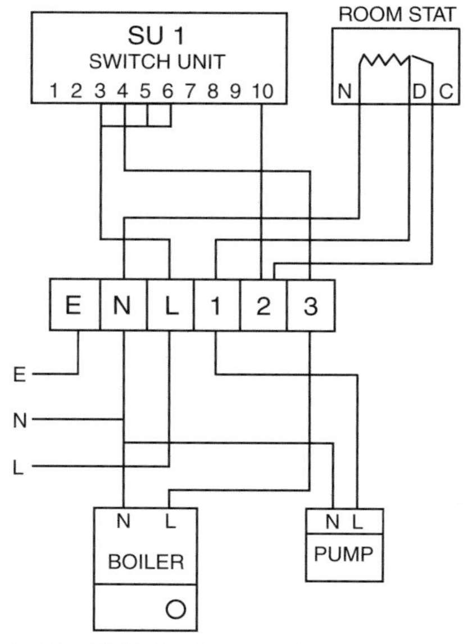

Figure 12.13 *Utilizing Drayton SU 1 switch unit*

Notes

a) Could be used with time clock.
b) Connect frost thermostat across junction box terminals L–3 for hot water or double-pole frost thermostat L–3 and L–1.

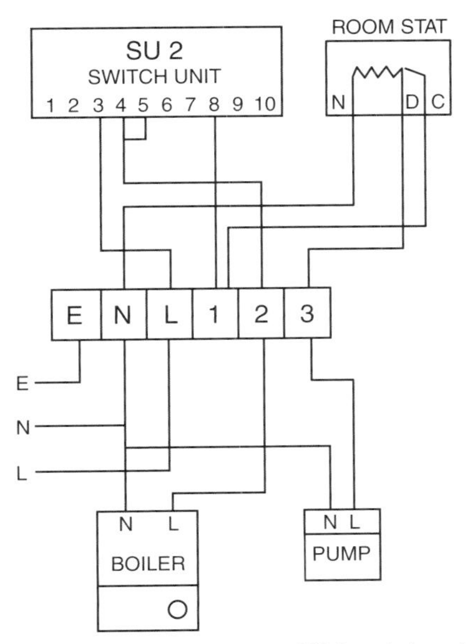

Figure 12.14 *Utilizing Drayton SU 2 switch unit*

Notes

a) Could be used with time clock.
b) Connect frost thermostat across junction box terminals L–2 for hot water or double-pole frost thermostat L–2 and L–3.

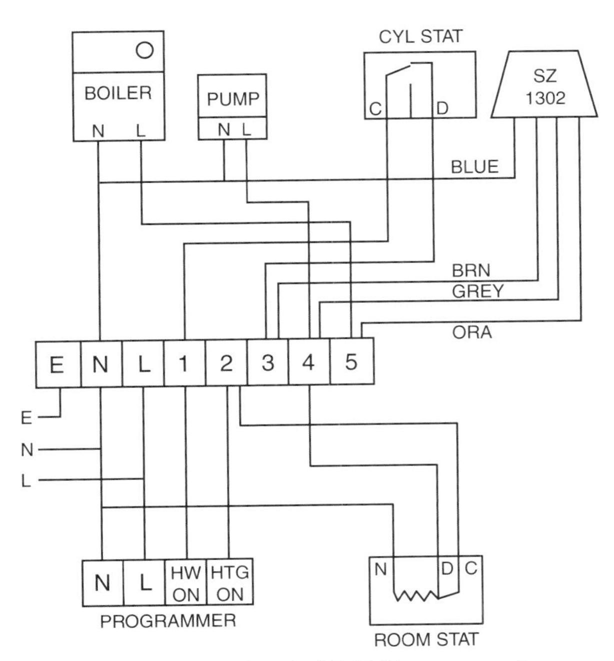

Figure 12.15 *With Sunvic SZ 1302 actuator to hot water circuit, giving full temperature and programming control*

Notes

a) Suitable for Basic or Full programming.
b) Connect frost thermostat across junction box terminals L–4.

Fully pumped systems

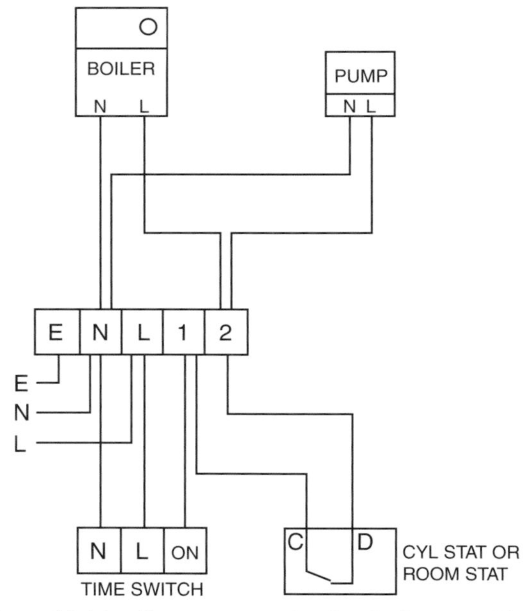

Notes

a) Where no room/cylinder thermostat exists, link 1–2 in junction box.
b) When using a pump overrun boiler then wire pump as indicated in Chapter 9.
c) Connect frost thermostat across junction box terminals L–1.

Figure 12.16 *For use on one circuit only, i.e. central heating or hot water, with or without temperature control*

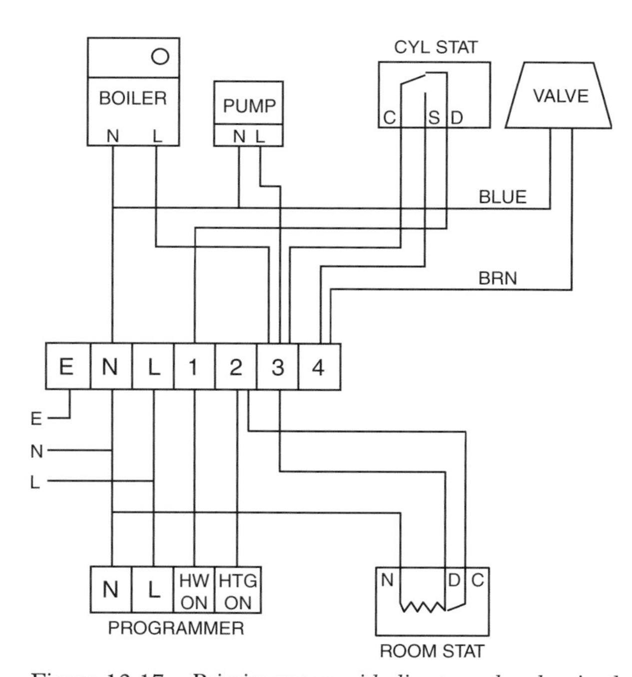

Notes

a) Suitable for Basic programmer only.
b) For specially manufactured priority programmers refer to page 58.
c) Connect frost thermostat across junction box terminals L–3.

Figure 12.17 *Priority system with diverter valve showing hot water priority*

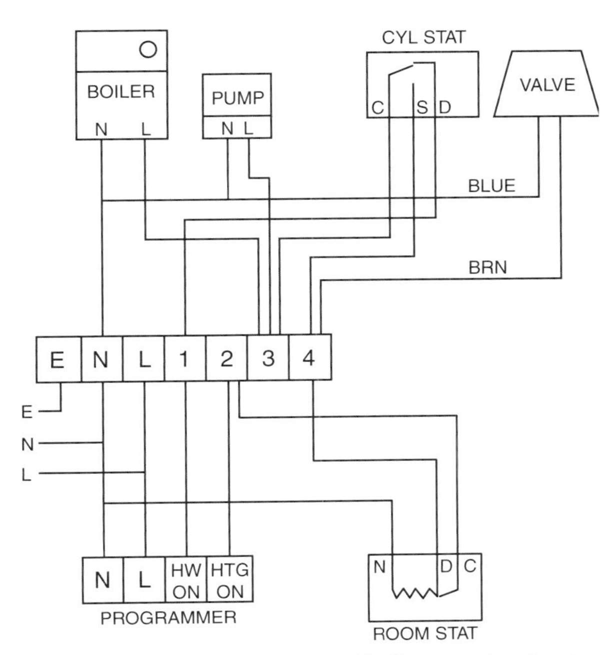

Figure 12.18 *Priority system with diverter valve showing central heating priority*

Notes

a) Suitable only for Basic programming.
b) For specially manufactured priority programmers refer to programmer section, page 58
c) Connect frost thermostat across junction box terminals L–3.

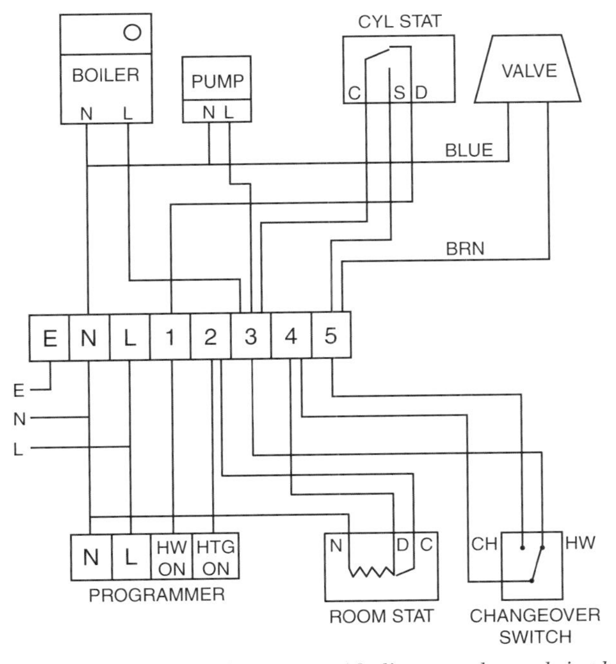

Figure 12.19 *Priority system with diverter valve and simple changeover switch to provide for optional priority*

Notes

a) Suitable for Basic programming only.
b) Connect frost thermostat across junction box terminals L–3.

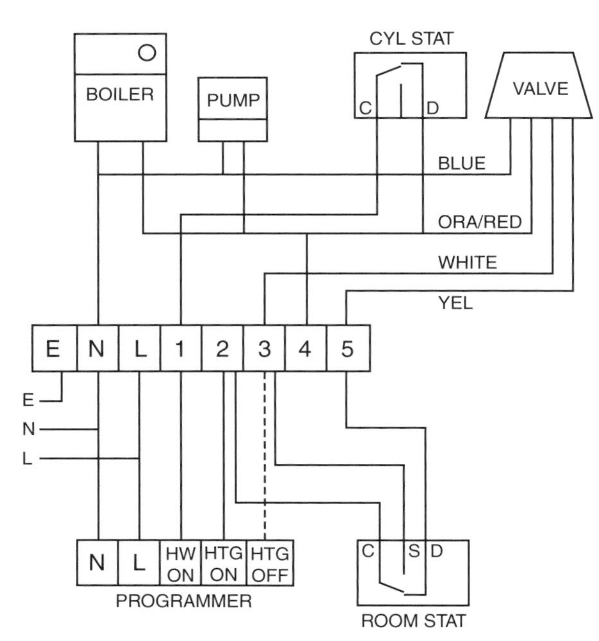

Notes

a) Hot water temperature control only when programmer is in 'Hot Water Only' position.
b) Programmer should have 'Heating Off' signal, otherwise room thermostat will need to be turned down whilst clock is in 'Heating On' mode.
c) System requires changeover room thermostat.
d) Suitable for Basic control only.
e) Connect frost thermostat across junction box terminals L–5.

Figure 12.20 *With one 2-port motor open, motor close motorized valve in central heating circuit*

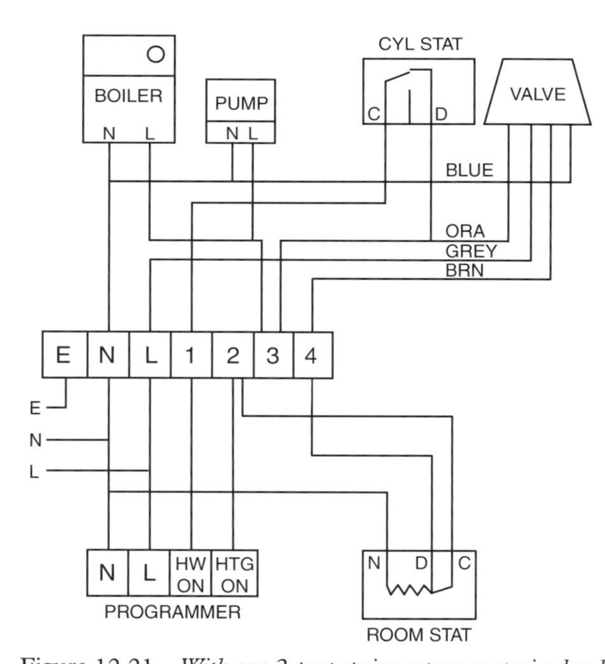

Notes

a) Suitable for Basic programming only.
b) Hot water temperature control only when programmer set to 'Hot Water' only.
c) Connect frost thermostat across junction box terminals L–4.

Figure 12.21 *With one 2-port spring return motorized valve in central heating circuit, providing limited hot water temperature control*

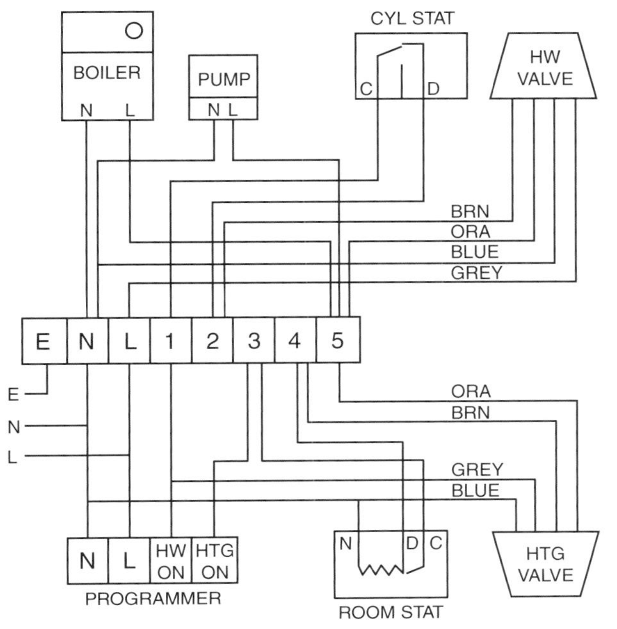

Figure 12.22 *With a 2-port spring return motorized valve to both central heating and hot water circuits, providing full temperature and programming control*

Notes

a) Known as Honeywell 'S' Plan and Sunvic System 4.
b) Suitable for Basic or Full programming.
c) Connect frost thermostat across junction box terminals L–4.

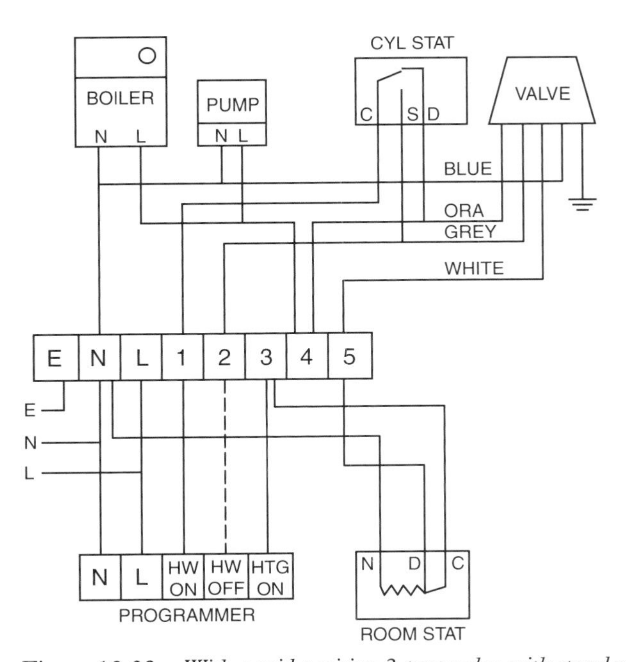

Figure 12.23 *With a mid-position 3-port valve with standard colour flex conductors. Provides full temperature and programming control*

Notes

a) Full programming only available if programmer has 'Hot Water Off' terminal and dotted connection is made.
b) Known as Honeywell 'Y' Plan and Sunvic Unishare System.
c) Connect frost thermostat across junction box terminals L–5.

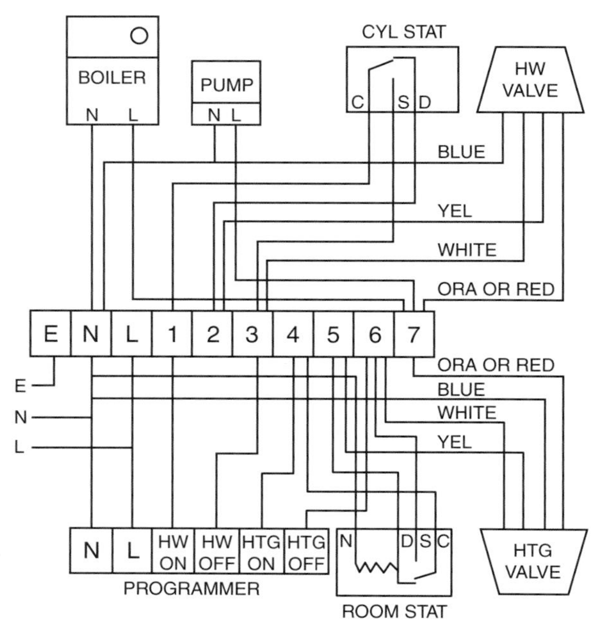

Notes

a) Programmer must have 'Offs' for both heating and hot water.
b) For frost protection refer to special diagrams.

Figure 12.24 *With a 2-port motor open, motor close motorized valve to both central heating and hot water circuits. Provides full temperature and programming control*

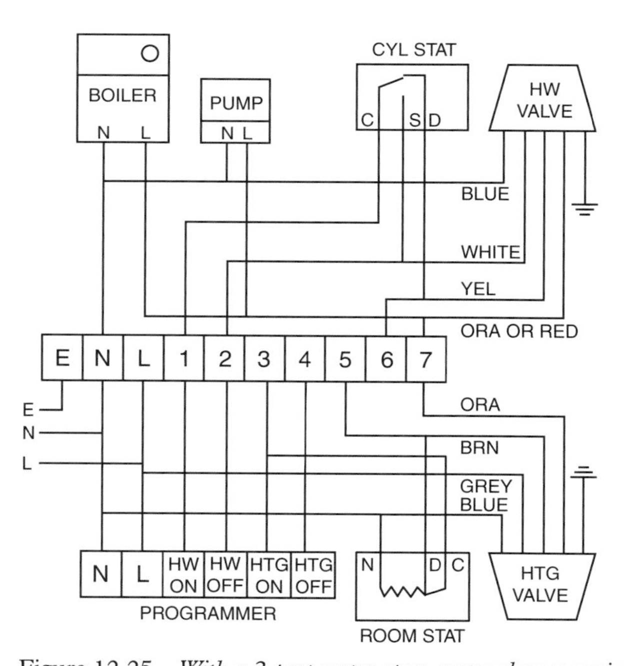

Notes

a) Suitable for Basic or Full control but programmer must have 'Off' signal for circuit with motor open, motor close valve (shown on hot water in this diagram).
b) Connect frost thermostat across junction box terminals L–5.

Figure 12.25 *With a 2-port motor open, motor close motorized valve and a 2-port spring return motorized valve to control central heating and hot water circuits*

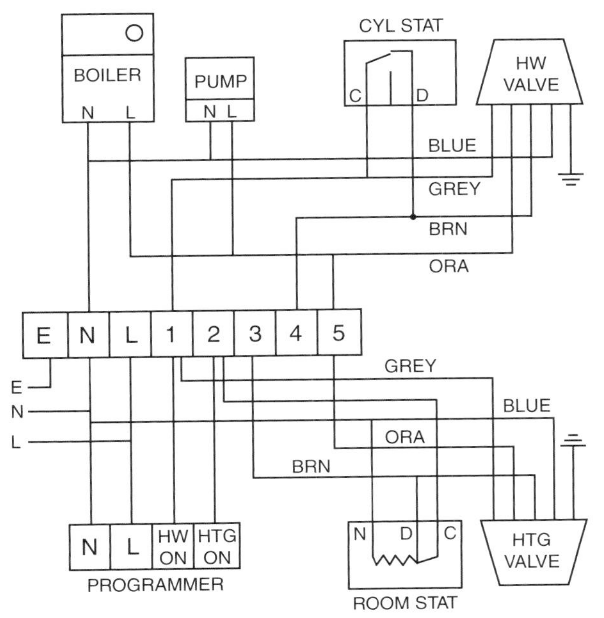

Figure 12.26 *With a 2-port spring return motorized valve to both central heating and hot water circuit when no permanent live is available for valve auxiliary switches. Provides full temperature control*

Notes

a) Suitable for Basic programming only.
b) Connect frost thermostat across junction box terminals L–3.

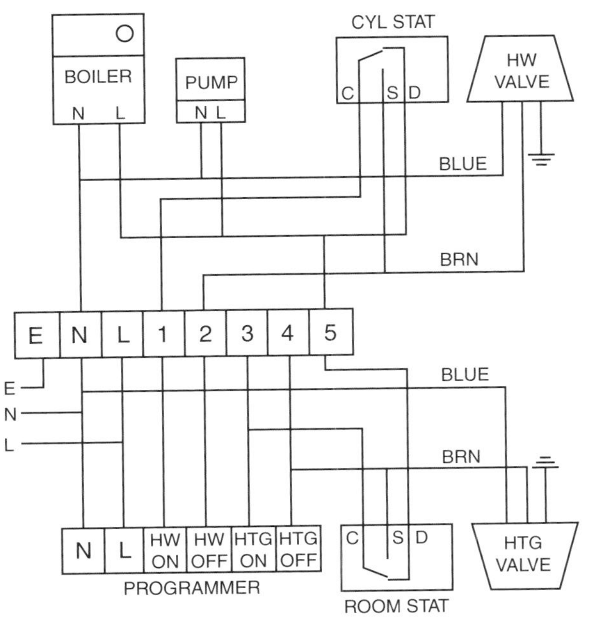

Figure 12.27 *With a 2-port spring return normally* open *motorized valve to both central heating and hot water circuits*

Notes

a) Known as Honeywell System G.
b) Not commonly used but type of valve V4043B used in solid-fuel systems.
c) Essential that programmer has 'Off' signals to both heating and hot water.
d) System requires changeover room and cylinder thermostat.
e) Connect frost thermostat across junction box terminals L–5.

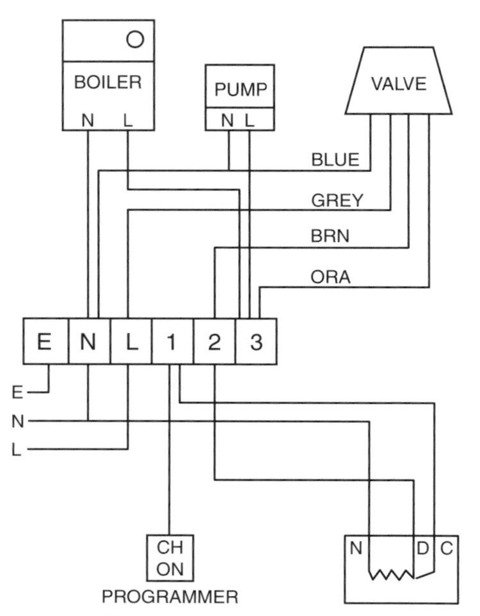

Note
Shown on central heating but could apply to the hot water circuit if required.

Figure 12.28 *Showing the addition of a 2-port spring return motorized valve to an existing system for zone control*

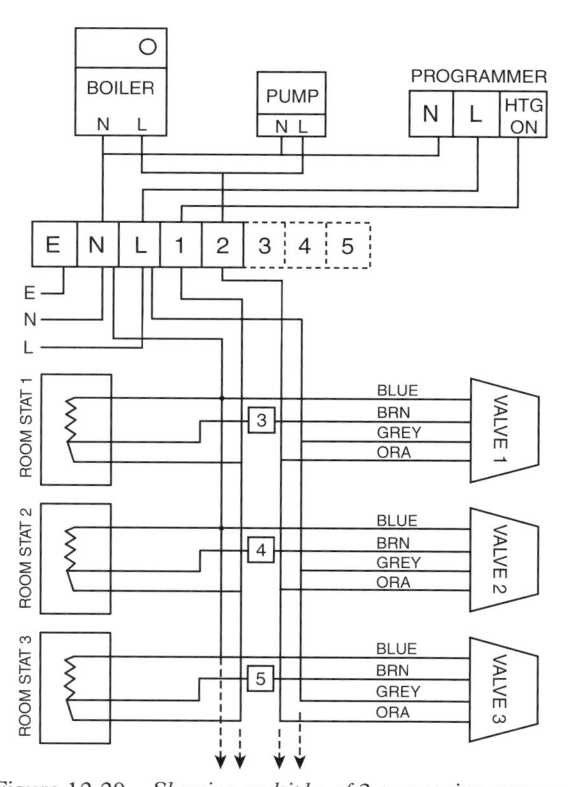

Note
Up to three valves are shown on heating, although in theory there is no limit and they could also be used in the hot water circuit, where several cylinders are utilized.

Figure 12.29 *Showing multiples of 2-port spring return motorized valves wired to one programmer, as may be used in a large property for zone control*

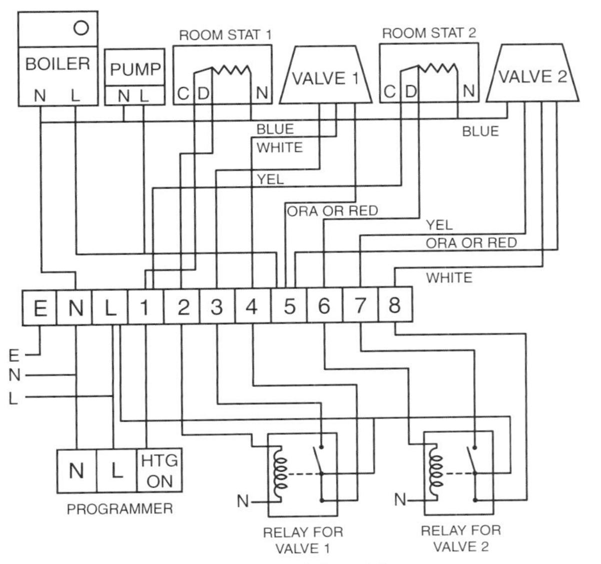

Notes

a) Diagram shows two valves on the heating, although in theory any number can be used, providing a relay is used for each. Note that the 'Heating Off' signal from the programmer is not required. The same principle applies if used on the hot water circuit.

b) In a system utilizing, e.g. one motor open, motor close valve on the hot water and two motor open, motor close valves on the heating, the hot water valve would be wired as normal using the 'Hot Water Off' from programmer and cylinder thermostat, and the heating valves should be wired as above. However, if there were two motor open, motor close valves to both hot water and heating circuits then all four would need to be wired as shown, and four relays would be required. As this would normally occur in a large property, it would be easier and more cost effective, and provide greater benefits in programming to fit two programmers or change actuators to spring return type. Remember that where a valve is controlled by a relay, a changeover thermostat or 'Off' signal from the programmer is not required.

Figure 12.30 *Showing multiples of 2-port motor open, motor close motorized valves wired to one programmer, as may be used in a large property for zone control*

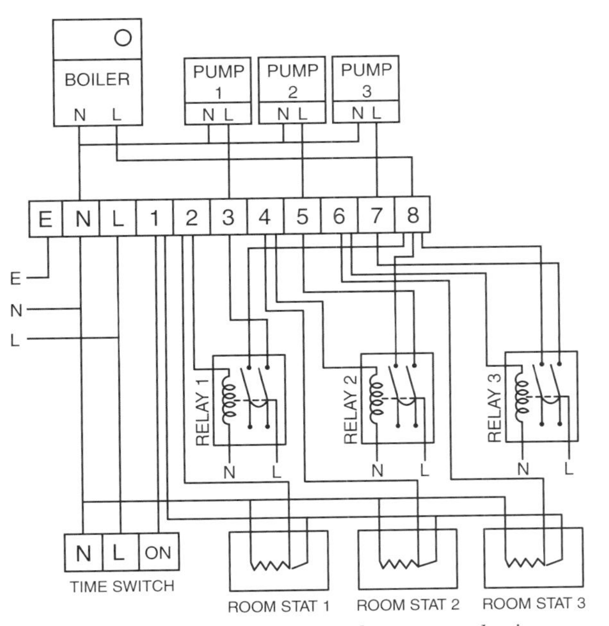

Notes

a) Requires a DPDT relay for each pump.

b) More appropriate to a commercial or large domestic situation.

c) Connect frost thermostat across junction box terminals L–1. It is assumed that at least one room thermostat is calling for heat.

Figure 12.31 *With one boiler and zone control using one pump for each zone. No motorized valves*

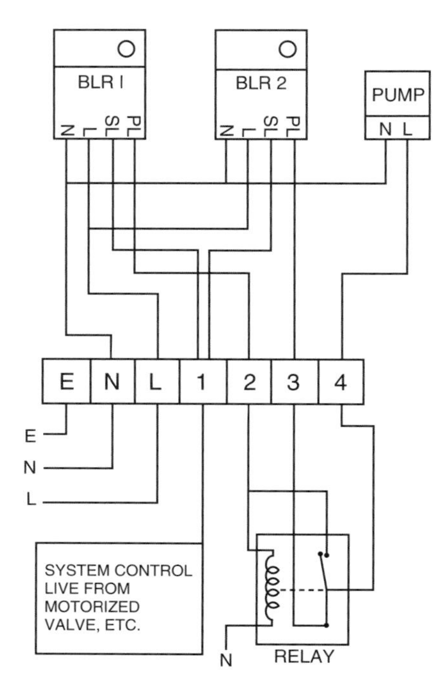

Notes

a) Requires a single-pole changeover relay for each boiler after the first one, e.g. two boilers, one relay, three boilers, two relays, etc.
b) Boiler PL refers to pump live. Boiler SL refers to switched live from system control.
c) Connect frost thermostat across junction box terminals L–2.

Figure 12.32 *With two or more pump overrun boilers and one pump using relays as required*

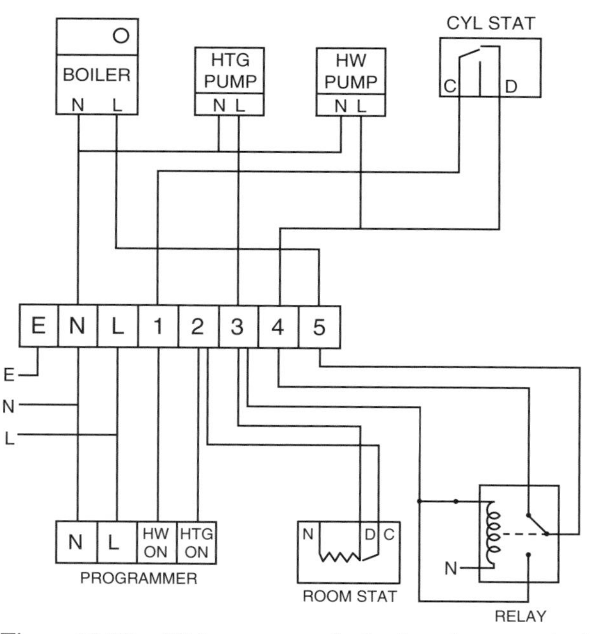

Notes

a) Requires a single-pole changeover relay (SPDT).
b) System similar to SMC control pack MK1 and can be utilized to convert from original Horstmann programmer.
c) Basic or Full control, dependent on programmer used.
d) Connect frost thermostat across junction box terminals L–3.

Figure 12.33 *With pump to each circuit and no motorized valves*

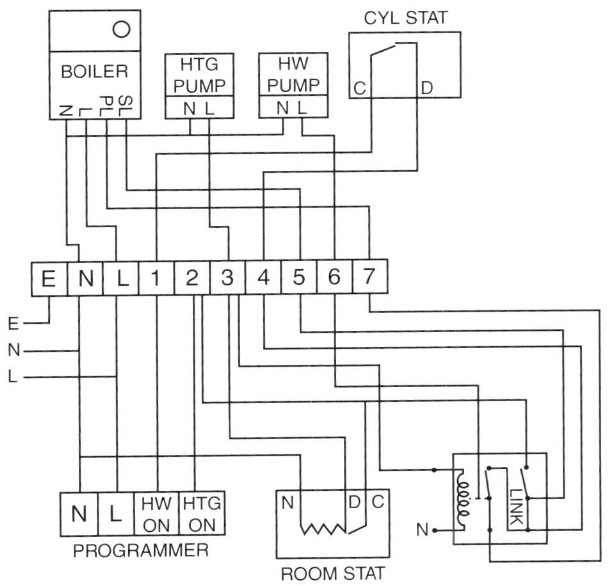

Notes

a) Requires a double-pole changeover relay (DPDT).
b) System similar to SMC control pack MK1 and can be utilized to convert from original Horstmann programmer.
c) Basic or Full control dependent on programmer used.
d) Connect frost thermostat across junction box terminals L–3.

Figure 12.34 *With pump overrun boiler, pump to each circuit, room and cylinder thermostats, programmer and relay. No motorized valves*

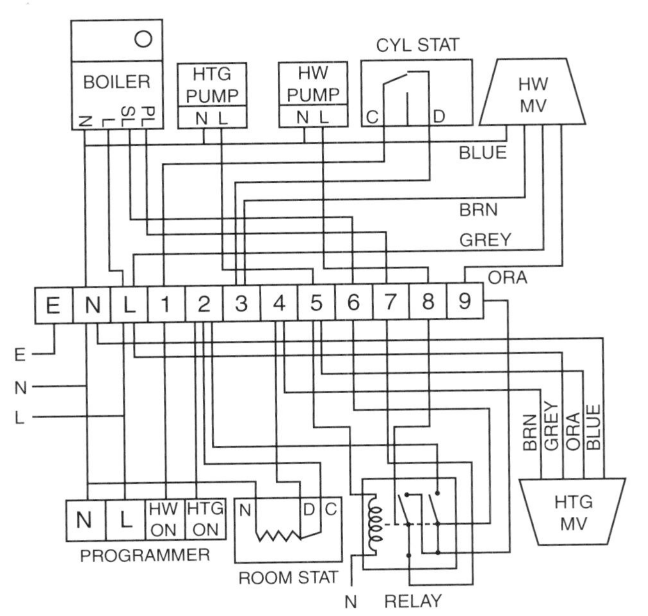

Notes

a) Requires a double-pole changeover relay (DPDT).
b) Pump overrun functions on hot water pump.
c) Connect frost thermostat across junction box terminals L–4.

Figure 12.35 *With pump overrun boiler, pump and 2-port spring return motorized valve to each circuit, room and cylinder thermostats, programmer and relay*

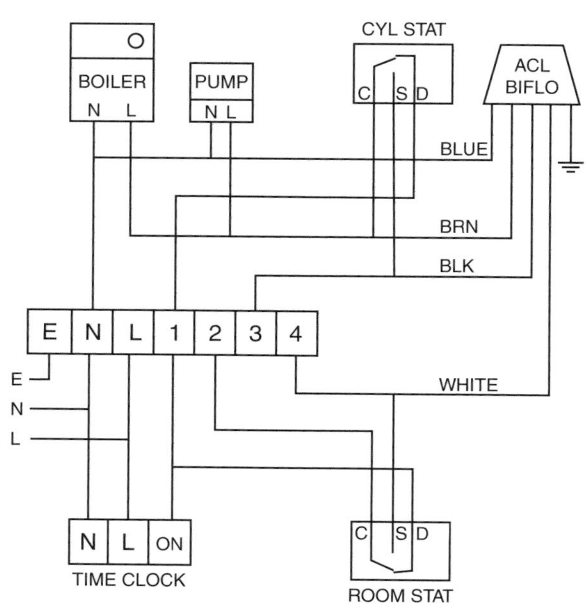

Figure 12.36 *ACL Biflo System MK1 with time clock*

Notes

a) System requires changeover room and cylinder thermostats.
b) See motorized valve section (Chapter 6, ACL 672 BRO 340).
c) Connect frost thermostat across junction box terminals L–1.

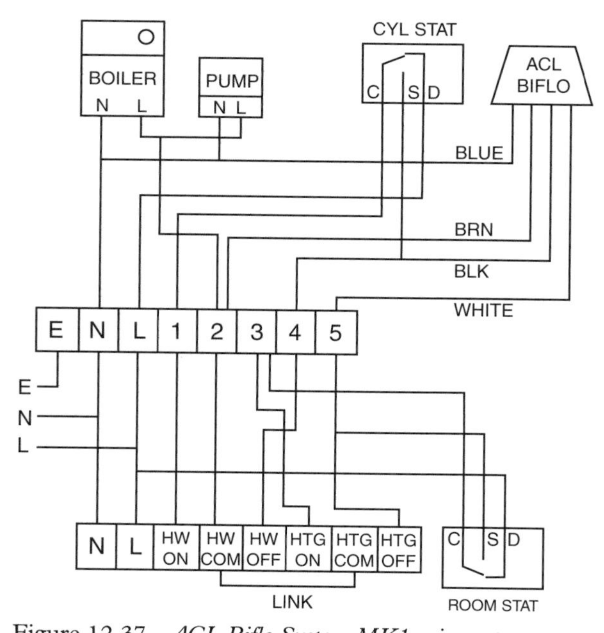

Figure 12.37 *ACL Biflo System MK1 using a programmer with voltage free contacts. Provides full temperature and programming control*

Notes

a) Programmer must have voltage free contacts – link as shown.
b) Brown wire of valve may be red.
c) System requires changeover room and cylinder thermostats.
d) Connect frost thermostat across junction box terminals L–2.

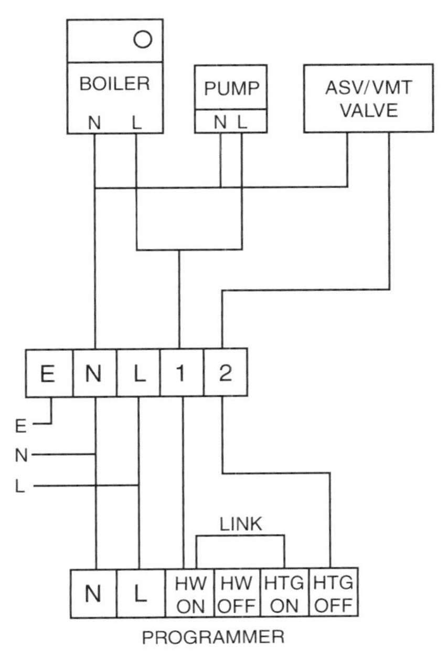

Figure 12.38 *Danfoss 2.2 and 2C system*

Notes

a) Suitable for Basic programming only.
b) Connect frost thermostat across junction box terminals L–1.

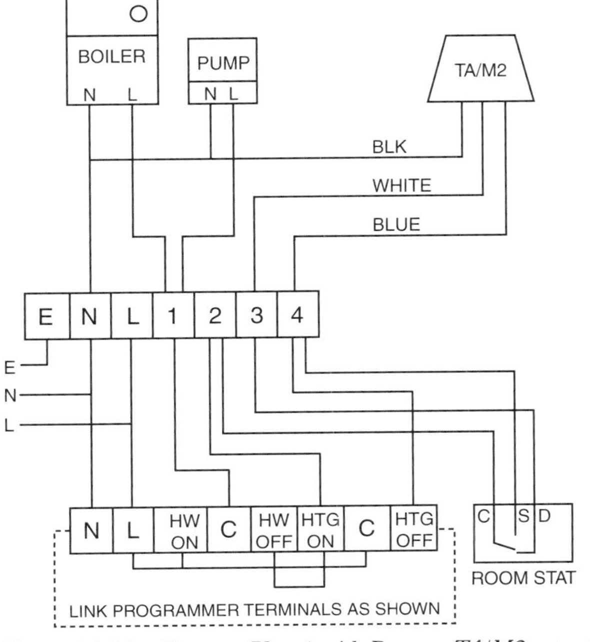

Figure 12.39 *Drayton Plan 1 with Drayton TA/M2 actuator*

Notes

a) No temperature control of hot water unless non-electric device is used.
b) Suitable for Basic programmer only, although programmer must have hot water and heating 'Offs' and voltage free switching.
c) Note that valve black is neutral.
d) Connect frost thermostat across junction box terminals L–3.

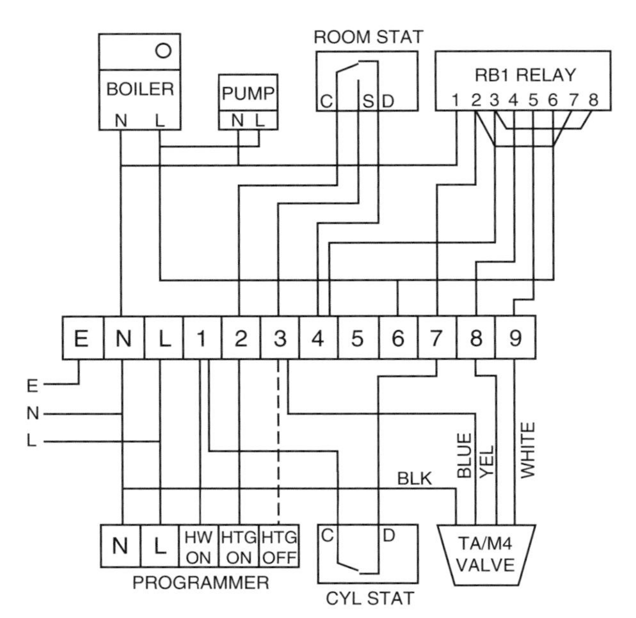

Notes

a) Suitable for Basic programmers only unless programmer has 'Heating Off' facility and dotted line is connected.
b) If a time switch is used link 'Heating On' and 'Hot Water On'.
c) As this actuator is reversible, it is possible to find different site wiring to that shown.
d) Relay diagram is inside cover. Relay can be replaced by double-pole changeover relay (DPDT).
e) Note that neutral on the actuator is black.
f) Connect frost thermostat across junction box terminals L–6.

Figure 12.40 *Drayton Flowshare 5 System with Drayton TA/M4 actuator, RB1 relay box and programmer*

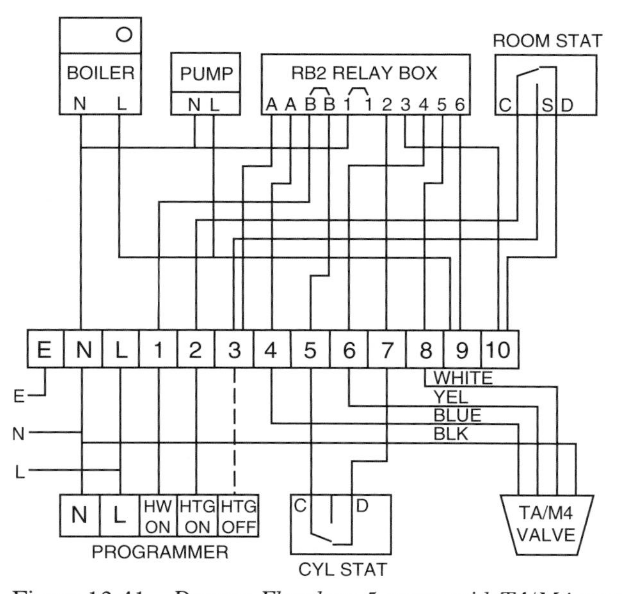

Notes

a) Suitable for Basic programmers only unless programmer has 'Heating Off' facility and dotted line is connected.
b) If a time switch is used link 'Heating On' and 'Hot Water On'.
c) As this actuator is reversible, it is possible to find different site wiring to that shown.
d) Relay diagram is inside cover. Relay can be replaced by double-pole changeover relay (DPDT).
e) Note that neutral on the actuator is black.
f) B–B and 1–1 are internally linked but A–A are not and must remain unlinked.
g) Connect frost thermostat across junction box terminals 1–6.

Figure 12.41 *Drayton Flowshare 5 system with TA/M4 actuator, RB2 relay box and programmer*

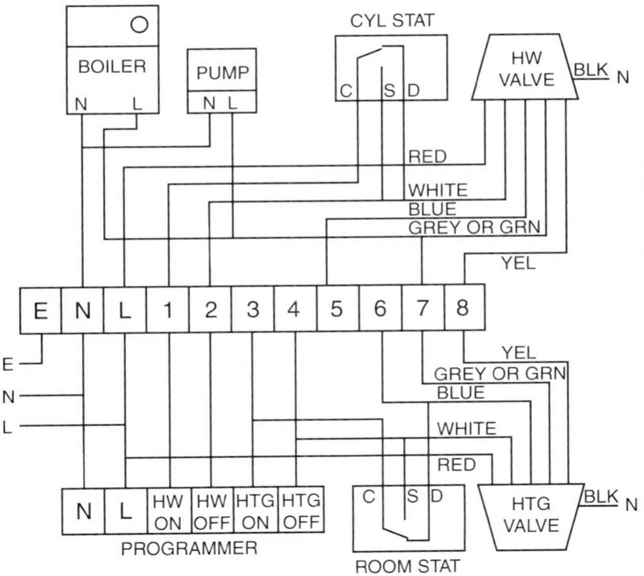

Notes

a) Essential that programmer has 'Off' signals to both heating and hot water.
b) System requires changeover room and cylinder thermostats.
c) Note that black is neutral and green is a live conductor.
d) Connect frost thermostat across junction box terminals L–6.

Figure 12.42 *Drayton Plan 7 System with two TA/M2A actuators providing full temperature and programming control*

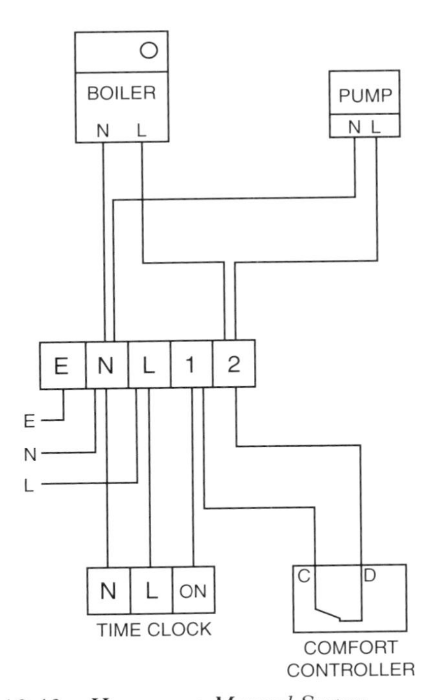

Notes

a) A manually operated 3-port valve is used to divert flow to heating or hot water circuits as required.
b) If no comfort controller, link 1–2 in junction box.
c) Connect frost thermostat across junction box terminals L–2.

Figure 12.43 *Homewarm Manual System*

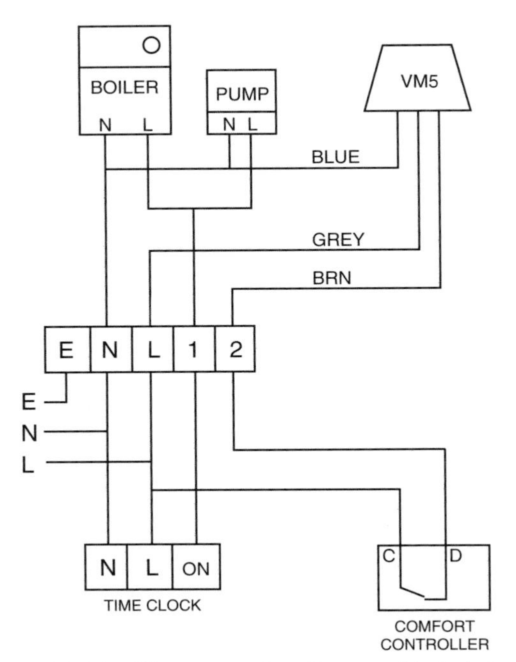

Figure 12.44 *Homewarm Auto System with Switchmaster VM5 actuator*

Notes

a) Designed as a low-cost installation. Refer to motorized valve section (Chapter 6, Switchmaster VM5).
b) If an orange wire is exposed in valve flex, this can be cut off and disregarded.
c) Connect frost thermostat across junction box terminals L–1.

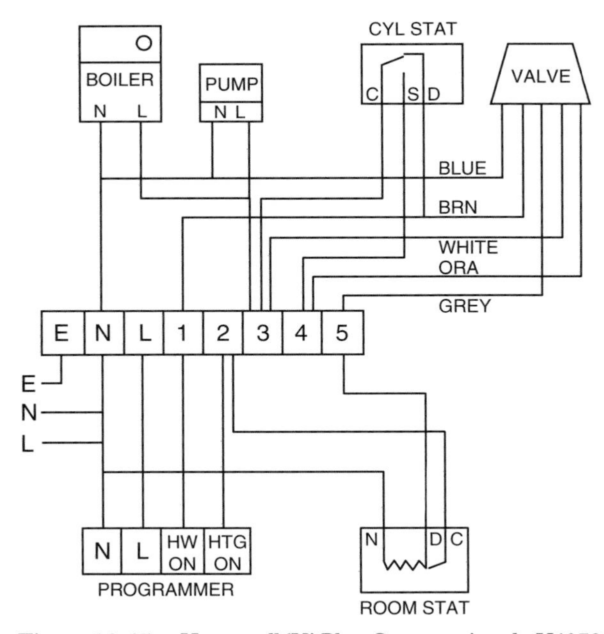

Figure 12.45 *Honeywell 'Y' Plan System using the V4073 6-wire, 3-port motorized valve with integral relay giving full temperature control*

Notes

a) Suitable in this diagram for a Basic programmer only.
b) Connect frost thermostat across junction box terminals L–3.

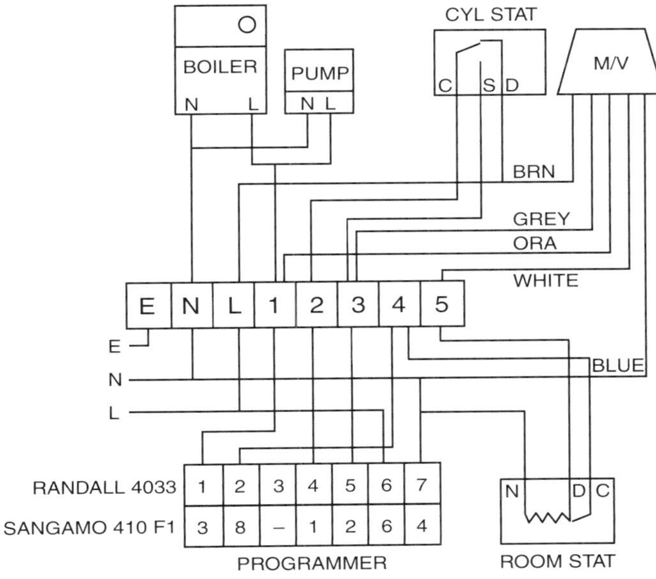

Figure 12.46 *Honeywell 'Y' Plan System using the V4073 6-wire, 3-port motorized valve with integral relay giving full temperature and programming control*

Notes

a) Suitable in this diagram for Full control using programmers shown.
b) For more information on the V4073 6-wire motorized valve refer to Chapter 6.
c) Connect frost thermostat across junction box terminals L–3.

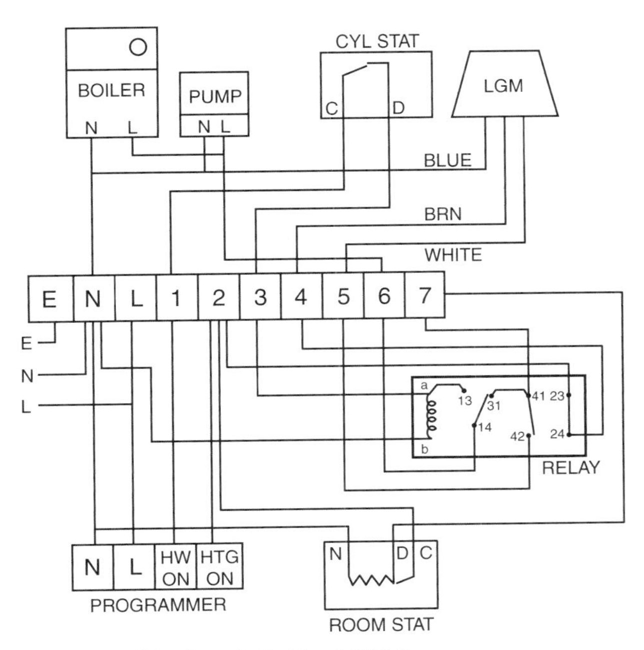

Figure 12.47 *Landis & Gyr LGM System*

Notes

a) Suitable for Basic or Full control, depending on programmer used.
b) Connect frost thermostat across junction box terminals L–7.

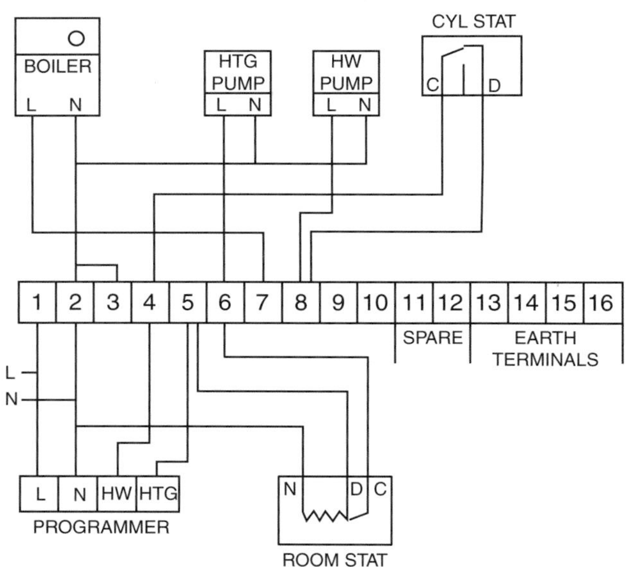

Notes

a) All controls wired into SMC wiring centre incorporating relay.
b) Internal wiring of SMC wiring centre not shown.
c) Suitable for Basic or Full programming.
d) Terminals 2–3 are linked internally for neutral connection.
e) Connect frost thermostat across junction box terminals L–6.

Figure 12.48 *SMC Control Pack 2 System with one boiler and a pump to both central heating and hot water circuits. No motorized valves. Provides full temperature and programming control*

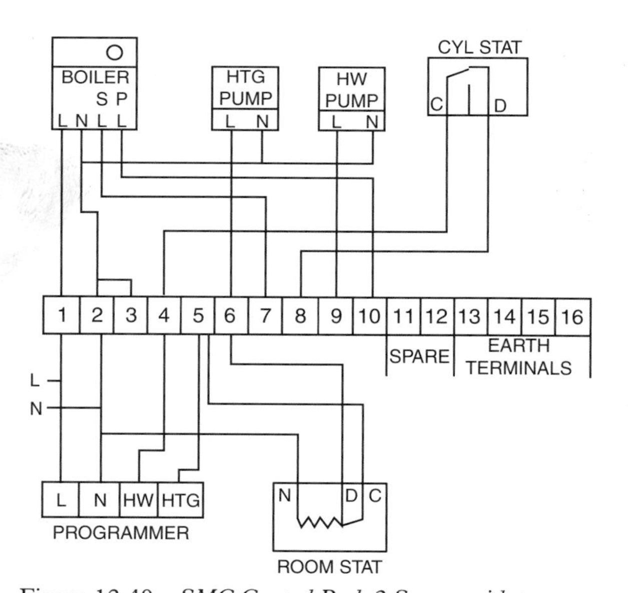

Notes

a) All controls wired into SMC wiring centre incorporating relay.
b) Internal wiring of SMC wiring centre not shown.
c) Suitable for Basic or Full programming.
d) Terminals 2–3 are linked internally for neutral connection.
e) Connect frost thermostat across junction box terminals L–6.

Figure 12.49 *SMC Control Pack 2 System with pump overrun boiler and pump to both central heating and hot water circuits. No motorized valves. Provides full temperature and programming control*

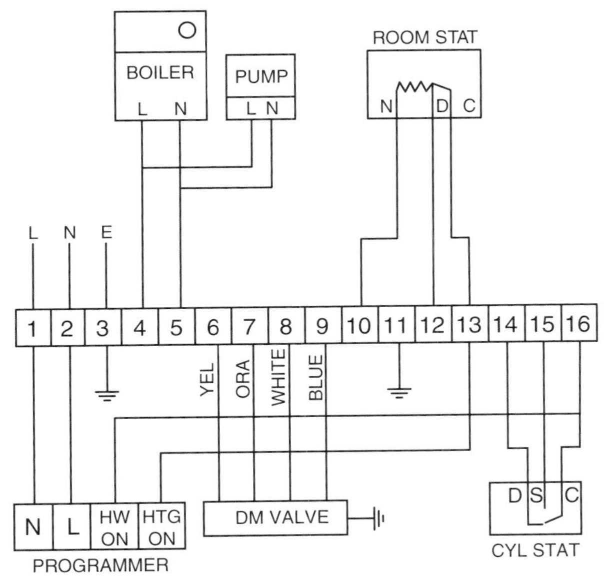

Figure 12.50 *Sunvic Duoflow System with RJ 1801 relay box and programmer*

Notes

a) Suitable for Basic programming only.
b) Yellow CH/HW, orange HW, white CH, blue N.
c) Can be used with any DM actuator.
d) Where a programmer is used, remove links 1–13 and 13–16.
e) Where a time switch is used, remove links 13–16.
f) Where there is no time control leave links in place.
g) Connect frost thermostat across junction box terminals 1–12.

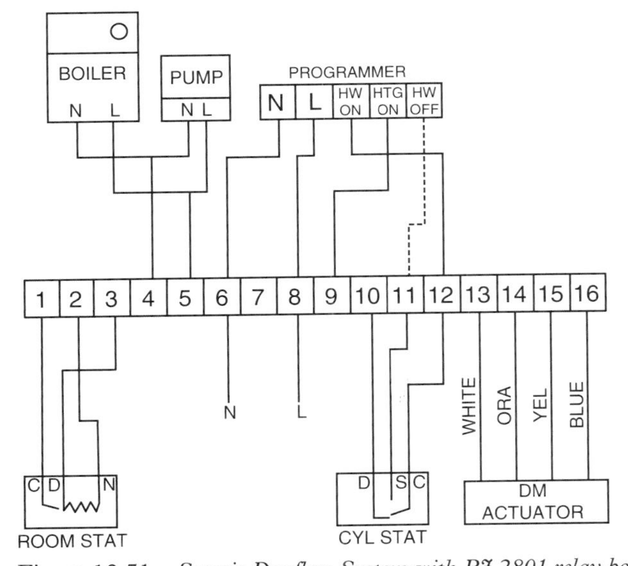

Figure 12.51 *Sunvic Duoflow System with RJ 2801 relay box and programmer*

Notes

a) Suitable for Basic programmer only unless programmer has ‘Central Heating Only’ and ‘Heating Off’ signal, in which case dotted line must be connected.
b) Can be used with any DM actuator.
c) Connect frost thermostat across junction box terminals 3–8.

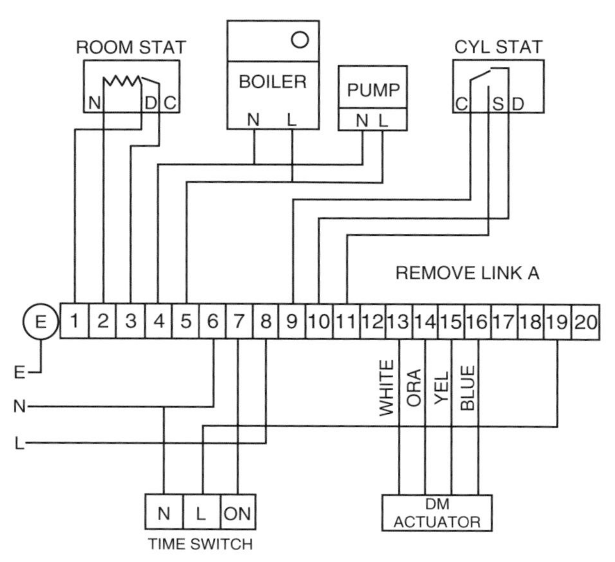

Notes

a) Remove link A.
b) Can be used with any of the DM actuators.
c) RJ 2852 is the plug-in version of the RJ 2802.
d) Connect frost thermostat across junction box terminals 1–19.

Figure 12.52 *Sunvic Duoflow System with RJ 2802 or RJ 2852 relay box and time clock*

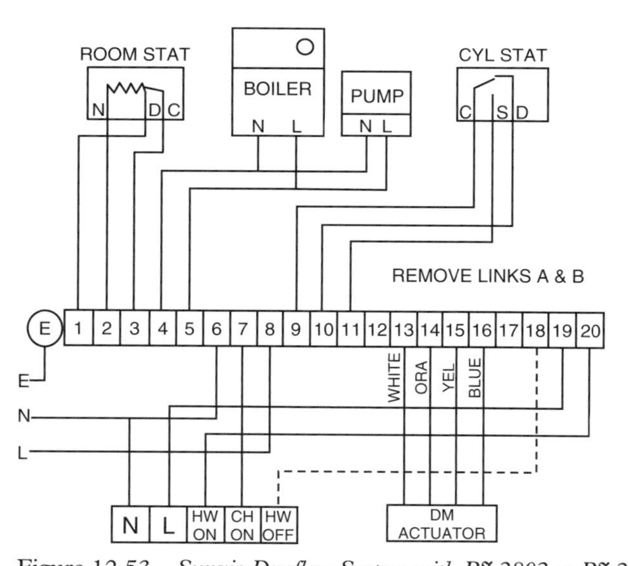

Notes

a) Remove links A and B.
b) Suitable for Basic programming only unless programmer has 'Central Heating Only' and 'Heating Off' signal, in which case dotted line must be connected.
c) Can be used with any of the DM actuators.
d) RJ 2852 is the plug-in version of the RJ 2802.
e) Connect frost thermostat across junction box terminals 1–19.

Figure 12.53 *Sunvic Duoflow System with RJ 2802 or RJ 2852 relay box and programmer*

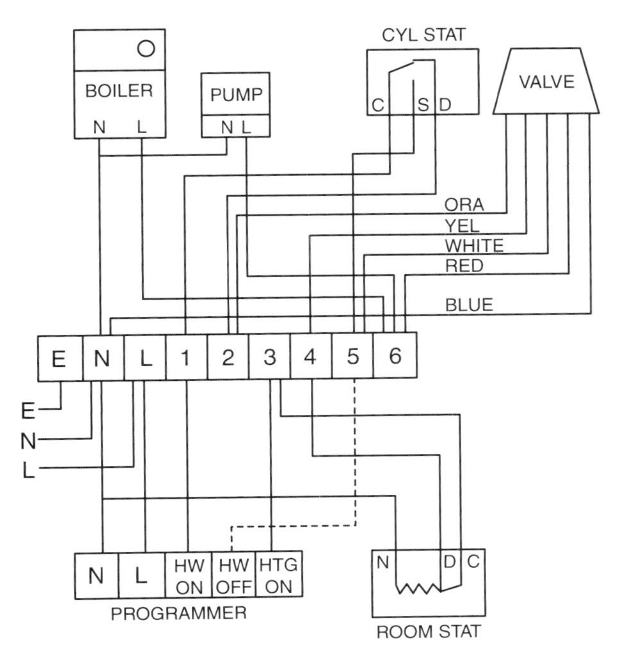

Figure 12.54 *Switchmaster Midi System*

Notes

a) Suitable for Basic control only unless programmer has 'Hot Water Off' signal, in which case full control is possible by connecting dotted line.

b) Connect frost thermostat across junction box terminals L–4.

Various frost protection thermostat wiring

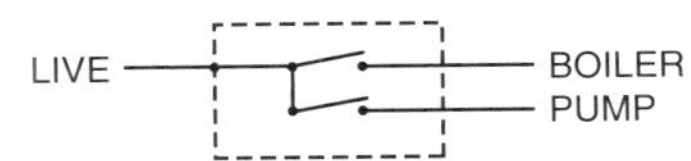

Figure 12.55 *Wiring of a double-pole frost thermostat to a gravity hot water, pumped central heating system*

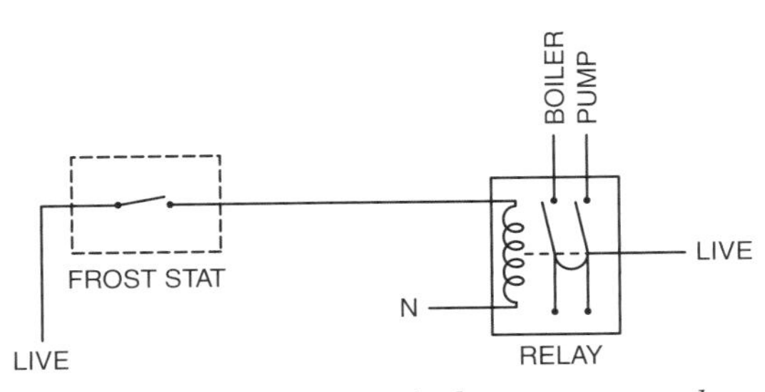

Figure 12.56 *Wiring of a single-pole frost thermostat to a gravity hot water, pumped central heating system using a DPDT relay*

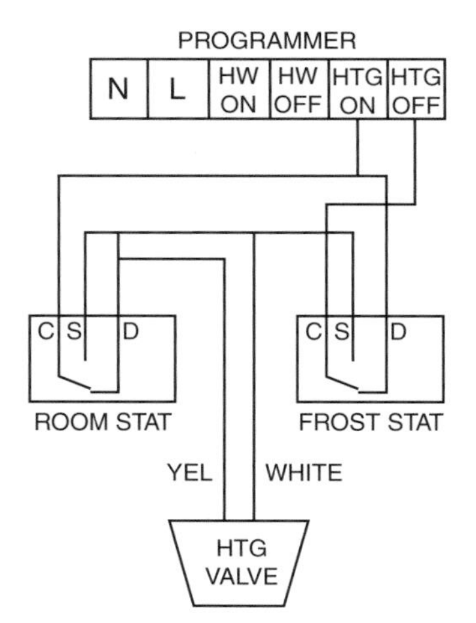

Notes

a) Will only function if room thermostat is also calling for heat.
b) To connect into existing, remove wire from programmer 'Heating Off' to room thermostat 'Satisfied'. Wire from programmer 'Heating Off' to frost thermostat 'Common'. Wire frost thermostat 'Satisfied' to room thermostat 'Satisfied' and wire frost thermostat 'Demand' to room thermostat 'Common'.
c) Requires SPDT room and frost thermostats.

Figure 12.57 *Wiring a frost thermostat to a motor open, motor close motorized valve*

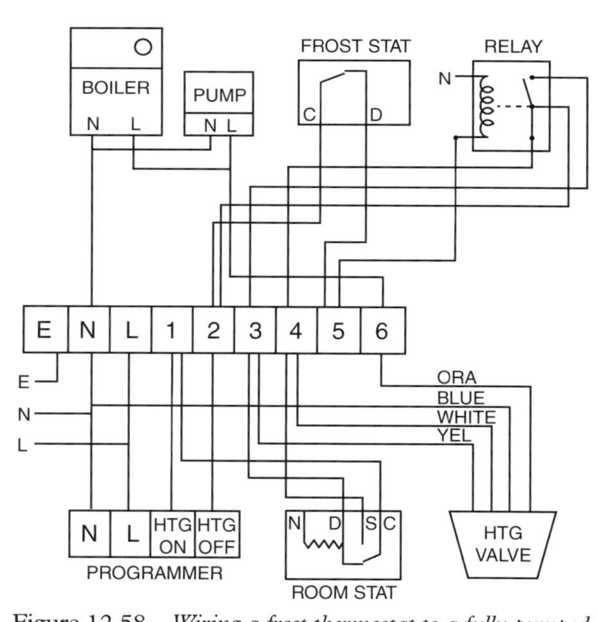

Notes

a) Frost thermostat shown on heating circuit but can be used on hot water if required.
b) Requires an SPDT replay.
c) Requires an SPST frost thermostat.
d) Frost thermostat common can go to terminal L if required instead of 'Heating Off'.

Figure 12.58 *Wiring a frost thermostat to a fully pumped system using two motor open, motor close motorized valves*

Supplementary wiring diagrams

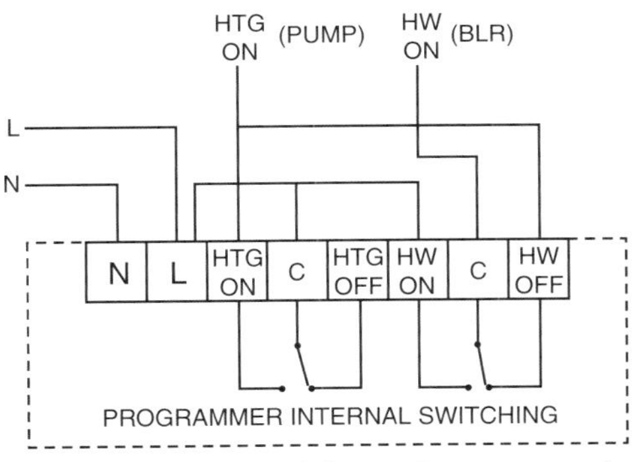

Figure 12.59 *Utilizing a full control programmer for basic control*

Note

Programmer must have voltage free terminals.

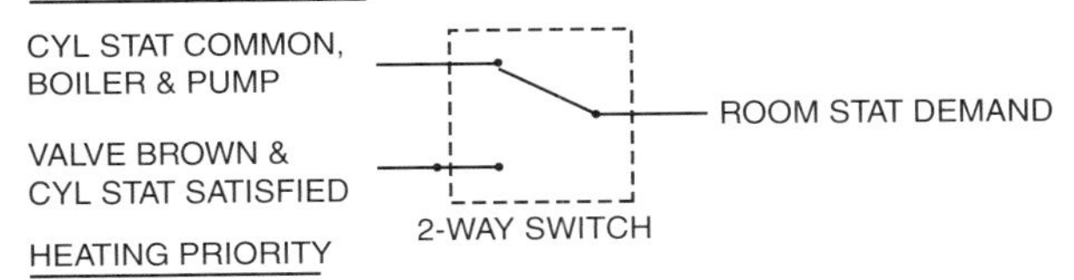

Figure 12.60 *Addition of a changeover switch to a priority system incorporating a 3-port valve, e.g. Honeywell V4044, to enable changing from hot water or heating priority as required. A simple 2-way light switch is ideal for this purpose.*

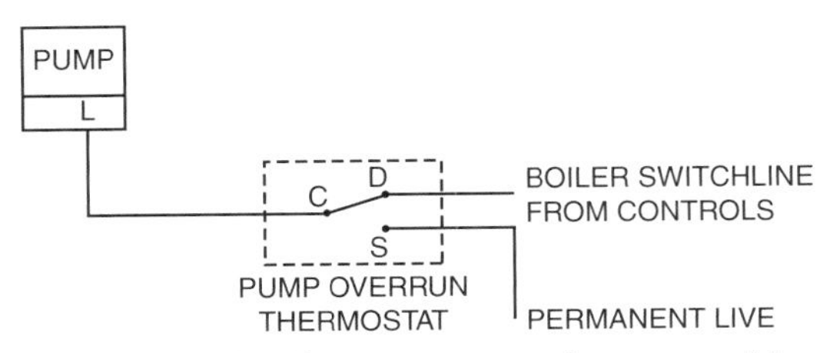

Figure 12.61 *Simple pump overrun thermostat wiring*

Note

Could be used as an additional control, e.g. on cast iron boilers, by employing a pipe thermostat located on the flow or return pipework as required.

Relays

Relays are often viewed with suspicion and a lot of electricians will hold their hands up in horror at the sight of a relay, especially when included into what already looks a complicated heating system. Relays deployed in this way are usually for switching 240 V to get over problems of 'back-feed', which may cause the system to do strange things. They can also be used for switching different voltages, e.g. a 240 V coil could switch 24 V and vice versa. There are a number of diagrams in the book which require the use of a relay and it may well be that the fault the engineer has been sent to find exists because a relay has never been fitted.

All relays are shown relaxed or de-energized.

ABBREVIATIONS:

SPDT	Single-pole double-throw – top
DPDT	Double-pole double-throw – middle
COM	Common
NO	Normally open – de-energized
NC	Normally closed – de-energized

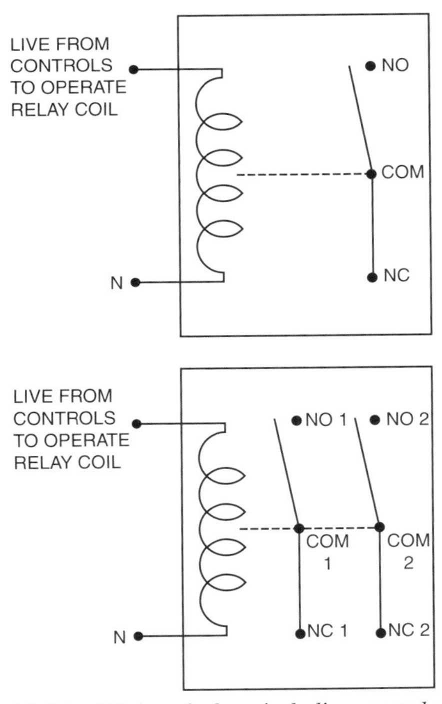

Figure 12.62 *Wiring of relays, including examples*

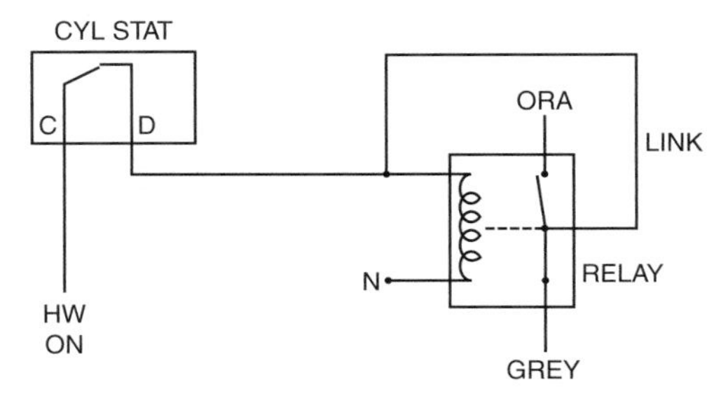

Notes

a) Ideal for when two wires have been run to a room or cylinder thermostat and system alterations require three wires.
b) Shown on a 3-way valve with standard flex conductors.

Figure 12.63 *Wiring of a relay to allow a 2-wire (SPST) cylinder or room thermostat to function as a 3-wire (SPDT)*

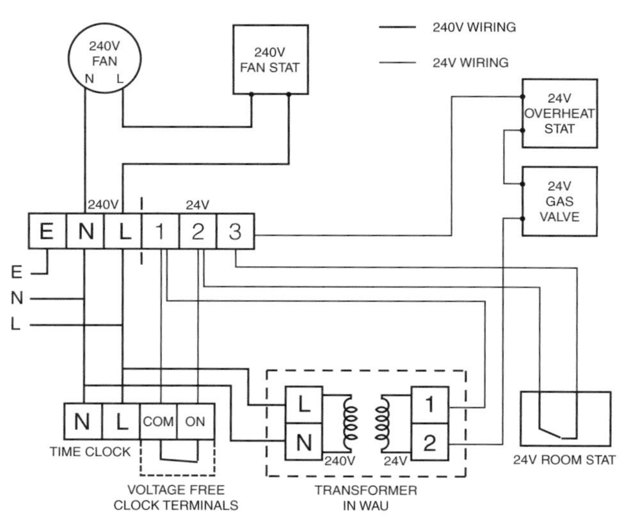

Notes

a) When the clock and room thermostat contacts are made, the main gas burner will light, warming the heat exchanger. The heat will be detected by the fan thermostat which will turn the fan on.
b) When the clock or room thermostat contacts break, the main gas valve will go off and the fan will continue to run for a while to clear the residual heat from the heat exchanger, until it falls below the setting of the fan thermostat.
c) It is usually essential that the time clock has voltage free terminals and in some clocks it is necessary to remove a link to achieve this.

Figure 12.64 *Simplified warm air unit wiring*

13

Interchangeability guide for programmers and time switches

This guide is to assist the replacement of faulty or obsolete programmers and time switches but it must be remembered that each group relates only to backplate and wiring, or product, and not to the facilities provided by the programmer or time switch itself, although a brief indication of setting options is given. Full details of the various differences, including dimensions, can be found earlier in this book.

This guide will also help in providing the householder with a prompt replacement, and in some cases offer the opportunity for replacing an electromechanical programmer with an electronic equivalent providing more programming options with minimum inconvenience. Obvious groupings are not included. For example there are four different models of Randall 103 time clock but are all wired the same and are interchangeable. Any differences are indicated earlier in this book.

The following groups of programmers and time switches can usually be directly interchanged without changing the backplate. However, backplate design may vary making a straight exchange impossible, e.g. a Landis & Gyr RWB2 will not fit an ACL 722 backplate but the reverse is possible. This is due to the provision of an earth terminal on the ACL backplate. Note also that it would be good working practice to change the backplate, as occasionally the contacts become worn and give rise to a fault situation that could mislead the engineer into suspecting that the programmer or time switch is faulty. In some cases it may be necessary to make a minor wiring alteration, and this is included in this chapter.

Programmers

Please note that all programmers listed below are suitable for use on Basic or Full systems unless stated otherwise. Therefore when changing ensure that the new programmer is set for the same system as the old one.

Group A

Electronic

ACL-Drayton Tempus 6 – 24 hour or 5/2 day programming
ACL-Drayton Tempus 7 (new) – 5/2 day or 7 day programming
ACL LP 112 – 24 hour programming
ACL LP 241 – 24 hour programming
ACL LP 522 – 5/2 day programming
ACL LP 722 – 7 day programming
ACL LS 112* – Basic systems only, 24 hour programming
ACL LS 241 – 24 hour programming
ACL LS 522 – 5/2 day programming
ACL LS 722 – 7 day programming
Barlo EPR 1 – 5/2 day programming
British Gas EMP 2
British Gas UP1

Danfoss Randall CP15 – 24 hour or 5/2 day programming
Danfoss Randall CP75 – 5/2 day or 7 day programming
Danfoss Randall FP15 – 24 hour or 5/2 day programming
Danfoss Randall FP75 – 5/2 day or 7 day programming
Danfoss Randall MP15 – Basic systems only, 24 hour or 5/2 day programming
Danfoss Randall MP75 – Basic systems only, 5/2 day or 7 day programming
Drayton Tempus 3 – 24 hour programming
Drayton Tempus 4 – 5/2 day programming
Drayton Tempus 7 – 7 day programming
Honeywell ST 6200 – Basic systems only, 24 hour programming
Honeywell ST 6300 – Full systems only, 24 hour programming
Honeywell ST 6400 – Full systems only, 7 day programming
Honeywell ST 6450 – Full systems only, 5/2 day programming
Horstmann Centaurplus C21 – 24 hour programming
Horstmann Centaurplus C27 – 7 day programming
Horstmann Centaurplus C121 – 24 hour programming
Horstmann Centaurplus C127 – 24 hour programming
Landis & Gyr RW 200 – 24 hour programming
Landis & Gyr RWB 20 – 7 day programming
Landis & Gyr RWB 40 – 24 hour programming
Landis & Gyr RWB 102 – Basic systems only, 24 hour programming
Landis & Gyr RWB 252 – 5/2 day programming
Landis & Steafa RWB 9 – 24 hour, 5/2 day, 7 day programming
Landis & Staefa RWB2E – 24 hour programming
Myson MEP 2C – 24 hour, 5/2 day, 7 day programming
Potterton Mini-minder E – 24 hour programming
Siemens RWB 29 – 24 hour, 5/2 day, 7 day programming
Siemens RWB 270 – 7 day programming
Smiths Centroller 1000 – 24 hour programming
Sunvic 207 – 24 hour, 5/2 day, 7 day programming
Tower DP 72 – 7 day programming
Tower QE2 – 7 day programming
* A wiring alteration will be necessary

Electromechanical

Crossling Controller
Glow-Worm Mastermind
Invensys SM 2 – 24 hour programming
Landis & Gyr RWB 1
Landis & Gyr RWB 2 MK 1
Landis & Gyr RWB 2 MK 2
Landis & Gyr RWB 2.9 – no neon indicators
Potterton Mini-minder
Tower SM 2 – 24 hour programming

Group B

Electronic

Danfoss Randall FP 965 – 5/2 day or 7 day programming
Horstmann 525 – 24 hour programming
Horstmann 527 – 7 day programming

Horstmann H21 – 24 hour programming
Horstmann H121 – 24 hour programming
Horstmann H27 – 7 day programming
Randall 922 – 24 hour programming
Randall 972 – 7 day programming
Randall Set 2** – Basic systems only, 24 hour programming
Randall Set 2E** – Basic systems only, 24 hour programming
Randall Set 3 – Full systems only, 24 hour programming
Randall Set 3E – 24 hour programming
Randall Set 5 – 5/2 day programming
**Link 1–5 on backplate.

Electromechanical

Danfoss 3002
Horstmann 425 Diadem
Horstmann 425 Tiara – no neon
Randall set 3M

Note: Only the Danfos Randall FP 965 will fit the Randall 922 and 972 backplate.

Group C

Electronic

Honeywell ST 499A – Full systems only, 24 hour programming
Honeywell ST 699B – Full systems only, 24 hour programming
Honeywell ST 699C – Full systems only, 24 hour programming
Thorn Microtimer – Full systems only, 24 hour programming

Group D

Electronic

Potterton EP 2000 – 24 hour programming
Potterton EP 2001 – 5/2 day programming
Potterton EP 2002 – 5/2 day programming
Potterton EP 3000 – 7 day programming
Potterton EP 3001 – 7 day programming
Potterton EP 3002 – 7 day programming
Potterton EP 6000 – 7 day and 5/2 day programming (optional each channel)
Potterton EP 6002 – Full systems only, 7 day programming

Group E

Electronic

Switchmaster 9000 – 24 hour programming
Switchmaster 9001 – 24 hour programming

Electromechanical

Switchmaster 900
Switchmaster 905

Time switches

Group A

Electronic

Horstmann H11 – 24 hour programming
Horstmann H17 – 7 day programming

Electromechanical

Danfoss 3001
Horstmann 425 Coronet

Group B

Electronic

Danfoss Randall TS975 – 5/2 day or 7 day programming
Randall 911 – 24 hour programming
Randall 971 – 7 day programming
Randall Set 1 – 1E – 24 hour programming
Randall Set 4 – 5/2 day programming
Sangamo Set 1 – 24 hour programming

Group C

See note after Group G.

Electronic

Horstmann Centaurplus C11 – 24 hour programming
Horstmann Centaurplus C17 – 7 day programming
Landis & Gyr RWB 50 – 24 hour programming
Landis & Gyr RWB 100 – 24 hour programming
Landis & Gyr RWB 152 – 5/2 day programming
Landis & Staefa RWB 7 – 24 hour, 5/2 day, 7 day programming
Landis & Staefa RWB 30E – 24 hour programming
Potterton Mini-minder ES – 24 hour programming
Siemens RWB 27 – 24 hour, 5/2 day, 7 day programming
Siemens RWB 170 – 7 day programming

Electromechanical

Landis & Gyr RWB 30

Group D

See note after Group G.

Electronic

ACL-Drayton Tempus 1 (new) – 24 hour or 5/2 day programming
ACL-Drayton Tempus 2 (new) – 5/2 day or 7 day programming
ACL LP 111 – 24 hour programming

ACL LP 711 – 7 day programming
ACL LS 111 – 24 hour programming
ACL LS 711 – 7 day programming
British Gas EMT 2
British Gas UT 1
Sunvic 107 – 24 hour, 5/2 day, 7 day programming

Electromechanical

Invensys SM1 – 24 hour programming

Group E

See note after Group G.

Electronic

Drayton Tempus 1 – 24 hour programming
Drayton Tempus 2 – 5/2 day programming

Group F

See note after Group G.

Electronic

Myson MEP 1c – 24 hour, 5/2 day, 7 day programming
Tower DT 71 – 7 day programming
Tower QE 1 – 7 day programming

Electromechanical

Tower QM1 – 24 hour programming

Group G

Electronic

Danfoss Randall TS 15 – 24 hour, 5/2 day programming
Danfoss Randall TS 75 – 5/2 day, 7 day programming
Honeywell ST 6100A – 24 hour programming
Honeywell ST 6100C – 7 day programming

Note: All time switches in Groups C, D, E, F and G have similar backplates but a wiring alteration will also be required.

Group H

Electronic

Potterton EP 4000 – 7 day programming
Potterton EP 4001 – 5/2 day programming
Potterton EP 4002 – 5/2 day programming
Potterton EP 5001 – 7 day programming
Potterton EP 5002 – 7 day programming

14

Manufacturers' trade names and directory

ACL Drayton	All enquiries to Invensys
Aga-Rayburn **Agaheat**	Station Road, Ketley, Telford, Shropshire, TF1 4DD Tel: 01952 642000; Fax: 01952 641961 info@aga-rayburn.co.uk www.aga-rayburn.co.uk
Alde Int. (UK) Ltd	Sandfield Close, Moulton Park, Northampton, NN3 1AB Tel: 01604 494193; Fax: 01604 499551 alde@aldeuk.force9.co.uk www.alde.co.uk
Alpha-Ocean Boilers	See Alphatherm Ltd
Alphatherm Ltd	United House, Goldsel Road, Swanley, Kent, BR8 8EX Tel: 01322 669443; Fax: 01322 615017 info@alphatherm.co.uk www.alphaboilers.com
Altecnic Ltd	Airfield Industrial Estate, Hixon, Staffs, ST18 0PF Tel: 01889 207200; Fax: 01889 270577 sales@altecnic.co.uk www.altecnic.co.uk
AMF-Venner	All enquiries to Heating Control Services
Amptec	See Electroheat
Andrews Water Heaters	Wednesbury One, Black Country New Road, Wednesbury, West Midlands, WS10 7NZ Tel: 0121 506 7400; Fax: 0121 506 7401 Andrews@rsawaterheating.co.uk www.andrewswaterheaters.co.uk
APECS Ltd	All enquiries to Quick Spares Ltd
Appliance Components Ltd (ACL)	All enquiries to Invensys
Applied Energy	Morley Way, Peterborough, PE2 9JJ Tel: 01733 456789; Fax: 01733 310606
Aquaflame Boilers	Unit 32, Haverscroft Industrial Estate, New Road, Attleborough, Norfolk, NR17 1YE Tel: 01953 454896; Fax: 01953 453792 sales@aquaflame.co.uk www.aquaflame.co.uk
Ariston Boilers	See MTS (GB) Ltd
Atag Heating	Unit M1, Hilton Business Park, East Wittering, West Sussex, PO20 8RL info@atagheating.co.uk www.atagheating.co.uk
Atlantic 2000 Boilers	PO Box 11, Ashton under Lyne, Lancashire, OL6 7TR Tel: 0161 621 5960; Fax: 0161 621 5966 info@atlantic2000.co.uk www.atlanticboilers.com
Atmos Heating Systems	West March, Daventry, Northants, NN11 4SA Tel: 01327 871990; Fax: 01327 871905 sales@atmos-heating.co.uk www.atmos-heating.co.uk
AWB	All enquiries to Time and Temperature

Barlo Products	Alexandra House, Block C, The Sweepstakes, Ballsbridge, Dublin 4, Ireland Tel: 00 353 231 0700; Fax: 00 353 231 0744 barlogroup@barlogroup.ie www.barlogroup.com
Baxi Potterton **Baxi Heating**	Brownedge Road, Bamber Bridge, Preston, Lancs, PR5 6SN Tel: 08706 060780; Fax: 01772 695410 info@baxi.co.uk www.baxi.co.uk
Biasi UK Ltd	Unit 31/33, Planetary Road Industrial Estate, Neachells Lane, Willenhall, Wolverhampton, WV13 3XB Tel: 01902 304400; Fax: 01902 304321 www.biasi.co.uk
B.H. Associates	3a & 3b Gilray Road, Vinces Road Industrial Estate, Diss, Norfolk, IP22 3EU Tel: 01379 640406; Fax: 01379 650640 Bha.merlin@fsbdial.co.uk www.merlinboilers.co.uk
Boulter Boilers Ltd	Magnet House, White House Road, Ipswich, IP1 5JA Tel: 01473 241555; Fax: 01473 241321 sales@boulter-boilers.com www.boulter-boilers.com
Brassware Ferroli	See Ferroli
Broag Limited	Remeha House, Molly Millars Lane, Wokingham, Berkshire, RG41 2QP Tel: 0118 978 3434; Fax: 0118 978 6977 boilers@broag-remeha.com www.broag-remeha.com
Burco Dean **Burco-Maxol**	Rose Grove, Burnley, Lancs, BB12 6AL Tel: 01282 427241; Fax: 01282 831206 sales@burcodean.co.uk www.burcodean.com
Caradon-Ideal Heating Ltd	PO Box 103, National Avenue, Hull, HU5 4JN Tel: 01482 492251; Fax: 01482 448858 enquiries@idealboilers.com www.idealboilers.com
R&S Cartwright	Floats Road, Roundthorn Industrial Estate, Manchester, M23 9NE
Calfire Boilers Ltd	See Turkington Engineering
Chaffoteaux et Maury	See MTS (GB) Ltd www.chaffoteaux.co.uk
Chalmor Ltd	Unit 1, Albert Road Industrial Estate, Luton Beds, LU1 3QF Tel: 01582 748700; Fax: 01582 748748 info@chalmor.co.uk www.chalmor.co.uk
Church Hill Systems	4A Hinckley Business Centre, Burbage Road, Hinckley, Leics, LE10 2TP Tel: 01455 890685; Fax: 01455 891341 sales@churchhillsystems.co.uk www.churchhillsystems.co.uk
Clyde Combustions	Cox Lane, Chessington, Surrey, KT9 1SL Tel: 020 8391 2020; Fax: 020 8397 4598 info@clyde4heat.co.uk www.clyde4heat.co.uk
Clyde Valley Control Systems Ltd	33 Glenburn Road, College Milton North, East Kilbride, G74 3BA Tel: 01355 247921; Fax: 01355 249197 sales@cvcs.co.uk www.storm-products.co.uk
Combat Engineering	Oxford Street, Bilston, West Midlands, WV14 7EG Tel: 01902 494425; Fax: 01902 403200 uksales@rg-inc.com www.rg-inc.com
Compact Electric Boilers	All enquires to Quick Spares Ltd
Crosslee PLC	Lightcliffe Factory, Hipperholme, Halifax, West Yorks, HX3 8DE Tel: 01422 203555; Fax: 01422 206304 lv@crosslee.co.uk www.crosslee.co.uk

Danfoss-Randall	Ampthill Road, Bedford, Beds, MK42 9ER Tel: 01234 364621; Fax: 01234 219705 danfossrandall@danfoss.com www.danfoss-randall.co.uk
Dimplex	Millbrook House, Grange Drive, Hedge End, Southampton, SO30 2DF Tel: 0870 077 7117; Fax: 0870 727 0109 www.dimplex.co.uk
Drayton Controls	All enquiries to Invensys
Dunphy Oil & Gas Burners Ltd	Queensway, Rochdale, Lancs, OL11 2SL Tel: 01706 649217; Fax: 01706 55512 sales@dunphy.co.uk www.dunphy.co.uk
Eastham Maxol	All enquiries to Burco Dean
Eberle Controls	8 Shannon Place, Potton, Sandy, Beds, SG19 2PZ Tel: 01895 444012; Fax: 01895 421901
Eco hometec (UK) Ltd	Unit 11E, Carcroft Enterprise Park, Carcroft, Doncaster, DN6 8DD Tel: 01302 722266; Fax: 01302 728634 sales@eco-homtec.co.uk www.eco-hometec.co.uk
Electroheat	3 Century House, Vickers Business Centre, Priestly Road, Basingstoke, Hants, RG24 9RA Tel: 01256 363417; Fax: 01256 841843 sales@electroheatplc.co.uk www.electroheatplc.co.uk
Elson	All enquiries to Elsy & Gibbons Ltd
Elsy & Gibbons Ltd	Simonside, South Shields, Tyne & Wear, NE34 9PE Tel: 0191 427 0777; Fax: 0191 427 0888 sales@elsonhotwater.co.uk www.elsonhotwater.co.uk
Euramo	See Wilo Salmson Pumps Ltd
Eurocombi	See MTS (GB) Ltd
Euro Controls	Unit 54, Monument Industrial Park, Chalgrove, Oxon, OX44 7RW Tel: 01865 400526; Fax: 01865 400524 sales@eurocontrols.com www.eurocontrols.com
Eurocal Boilers	See Turkington Engineering
Eurotronics	See Euro Controls
Fagor Comfort UK Ltd	Morco House, 59 Beverley Road, Hull, HU3 1XW Tel: 01482 325456; Fax: 01482 212869 sales@morcoproducts.co.uk
Ferroli	Lichfield Road, Branston Industrial Estate, Burton upon Trent Staffs, DE14 3HD Tel: 08707 282885; Fax: 08707 282886 sales@ferroli.co.uk www.ferroli.co.uk
Firebird Boilers	Unit 6, Westover Industrial Estate, Ivybridge, Devon, PL21 9ES Tel: 01752 691177; Fax: 01752 691131 info@firebirdboilers.co.uk www.firebirduk.co.uk
Firefly (UK) Ltd	Unit 4, Stag Business Park, Christchurch Road, Ringwood, Hants, BH24 3SB Tel: 01425 480210; Fax: 01425 479089
Flash	All enquiries to Euro Controls
GAH (Heating Products)	Melton Road, Melton, Woodbridge, Suffolk, IP12 1NH Tel: 01394 386699; Fax: 01394 386609 mail@gah.co.uk www.gahheatingproducts.co.uk

GEC Nightstor
All enquiries to Quick Spares Ltd

Gledhill Water Storage
Sycamore Estate, Squires Gate, Blackpool, Lancs, FY4 3RL
Tel: 01253 474444; Fax: 01253 474445
Sales@gledhill.net www.gledhill.net

Glow-Worm
See Hepworth Heating Ltd

GP Burners Ltd
2D Hargreaves Road, Groundwell Industrial Estate, Swindon, Wilts, SN2 5AZ
Tel: 01793 709050; Fax: 01793 709060
info@gpburners.co.uk www.gpburners.co.uk

Grant Engineering
Hopton House, Hopton Industrial Estate, Devizes, Wilts, SP2 7PU
Tel: 01380 736920; Fax: 0870 777 5553
info@grantuk.com www.grantuk.com

Grasslin (UK) Ltd
Vale Rise, Tonbridge, Kent, TN9 1TB
Tel: 01732 359888; Fax: 01732 354445
www.tfc-group.co.uk

Grundfos Pumps Ltd
Grovebury Road, Leighton Buzzard, Beds, LU7 4TL
Tel: 01525 850000; Fax: 01525 850011
www.grundfoss.com

Halstead Boilers
20/22 First Avenue, Bluebridge Industrial Estate, Halstead, Essex, CO9 2EX
Tel: 01787 475557; Fax: 01787 474588
sales@halsteadboilers.co.uk www.halsteadboilers.co.uk

Hamworthy Heating Ltd
Fleets Corner, Poole, Dorset, BH17 7LA
Tel: 01202 662500; Fax: 01202 665111
company.info@hamworthy-heating.com
www.hamworthy-heating.com

Heat Line
16–19 The Manton Centre, Manton Lane, Bedford, MK41 7PX
Tel: 0870 787 3363; Fax: 0870 777 8322
info@heatline.co.uk www.heatline.co.uk

Heating Control Services
Tel: 01922 634503; Fax: 01922 723777

Heating World Group
Excelsior Works, Eyre Street, Birmingham, B18 7AD
Tel: 0121 454 2244; Fax: 0121 454 4488
info@heatingworld.com www.heatingworld.com

Heatmiser UK Ltd
Primrose House, Primrose Street, Darwen, BB3 2DE
Tel: 01254 776343; Fax: 01254 704143
service@heatmiser.co.uk

Heatrae-Sadia
Hurricane Way, Norwich, Norfolk, NR6 6EA
Tel: 01603 420100; Fax: 01603 409409
sales@heatraesadia.com www.heatraesadia.com

Heb Boilers
All enquiries to GAH Heating Products

Hepworth Heating
Nottingham Road, Belper, Derbyshire, DE56 1JT
Tel: 01773 824141; Fax: 01773 828123
info@glow-worm.co.uk www.glow-worm.co.uk

Hermann
Broughton & Crangrove
Tel: 0870 6060601
hermann.uk@tiscali.co.uk www.hermann.it

Honeywell Control Systems
Honeywell House, Arlington Business Park, Bracknell, Berkshire, RG12 1EB
Tel: 01344 656000; Fax: 01344 656204
uk.infocentre@honeywell.com www.honeywell.com

Horstmann Controls	South Bristol Business Park, Roman Farm Road, Bristol, BS4 1UP Tel: 01179 788700; Fax: 01179 788701 sales@horstmann.co.uk www.horstmann.co.uk
Hoval Ltd	Northgate, Newark, Notts, NG24 1JN Tel: 01636 672711; Fax: 01636 673532 hoval@hoval.co.uk www.hoval@hoval.co.uk
HRM Boiler Co	Haverscroft Industrial Estate, Attleborough, Norfolk, NR17 1YE Tel: 01953 455400; Fax: 01953 454483 info@hrmboilers.co.uk www.hrmboilers.co.uk
Ideal	See Caradon-Ideal
IMI Pactrol	See Pactrol Controls
Imstor	See Church Hill Systems Ltd
Invensys	PO Box 57, Farnham Road, Slough, Berks, SL1 4UH Tel: 01753 550550; Fax: 01753 824078 Customer.care@invensys.com www.invensys.com
Jaguar	All enquiries to Hepworth Heating
JLB Group	All enquiries to Crosslee PLC
Johnson & Starley	Rhosili Road, Brackmills, Northampton, NN4 0LZ Tel: 01604 762881; Fax: 01604 767408 sales@johnsonandstarley.co.uk www.johnsonandstarley.co.uk
Keston Boilers	34 West Common Road, Hayes, Bromley, Kent, BR2 7BX Tel: 020 8462 0262; Fax: 020 8462 4459 info@keston.co.uk www.keston.co.uk
Landis & Gyr	All enquiries to Siemens
Landis & Staefa	All enquiries to Siemens
Lennox Industries	Cornwell Business Park, Slat House Road, Brackmills, Northampton, NN4 7EX Tel: 01604 669100; Fax: 01604 669150 www.lennoxuk.com
Lochnivar Ltd	7 Lombard Way, The MXL Centre, Banbury, Oxon, OX16 4TJ Tel: 01295 269981; Fax: 01295 271640 sales@lochnivar.lyd.uk www.lochnivar.ltd.uk
Malvern Boilers	Spring Lane North, Malvern, Worcs, WR14 1BW Tel: 01684 893777; Fax: 01684 893776 sales@malvernboilers.co.uk www.malvernboilers.co.uk
Maxol	All enquiries to Burco Dean
Merlin	See B.H. Associates
Meta	See Modular Heating Sales
Modular Heating Sales	35 Nobel Square, Burnt Mills Industrial Estate, Basildon, Essex, SS13 1LT Tel: 01268 591010; Fax: 01268 728202 sales@modular-heating-group.co.uk www.mhsboilers.com
MTS (GB) Ltd	MTS Building, Hughenden Avenue, High Wycombe, Bucks, HP13 5FT Tel: 01494 755600; Fax: 01494 459775 info@mtsgb.ltd.uk www.mtsgb.ltd.uk

Myson Controls	Eastern Valley, Team Valley, Gateshead, Tyne & Wear, NE11 0PG Tel: 0191 491 7530; Fax: 0191 491 7568 sales@myson.co.uk www.mysoncontrols.co.uk
Myson Heating	All enquiries to Baxi Potterton
Nailmere Ltd	All enquiries to Time & Temperature for AWB and Pyrocraft boilers
Nu-Heat Ltd	Heathpark House, Devonshire Road, Heathpark Industrial Estate, Honiton, Devon, EX14 1SD Tel: 01404 549770; Fax: 01404 549771 ufh@nu-heat.co.uk www.nu-heat.co.uk
Nu-Way Ltd	PO Box 1, Vines Lane, Droitwich, Worcs, WR9 8NA Tel: 01905 794331; Fax: 01905 794017 info@nu-way.co.uk www.nu-way.co.uk
Ocean Boilers	See Alphatherm Ltd
Offergram	Ceased trading. Spares from usual sources.
Ouzledale Foundry Co	Long Ing, Barnoldswick, Colne, Lancs, BB18 6BN Tel: 01282 813235; Fax: 01282 816876 esse@ouzledale.co.uk www.ouzledale.co.uk
Pactrol Controls	10 Pithey Place, West Pimbo, Skelmersdale, Lancs, WN8 9PS Tel: 01695 725152; Fax: 01695 724400 post@pactrol.com www.pactrol.com
Pakaway Perrymatics	Ceased trading. Spares from usual sources.
Paragon Electric Ltd	All enquiries to Heating Control Services
Parkray	All enquiries to Hepworth Heating Ltd
Pegler Ltd	St Catherines Avenue, Doncaster, DN4 8DF Tel: 0870 120 0284; Fax: 01302 560109 uk.sales@pegler.co.uk www.pegler.com
Perrymatic	Ceased trading. Spares from usual sources.
Potterton	All enquiries to Baxi Potterton enquiries@potterton.co.uk www.potterton.co.uk
Potterton Myson	All enquiries to Baxi Potterton
Powermatic	Winterhay Lane, Ilminster, Somerset, TA19 9PQ Tel: 01460 53535; Fax: 01460 52341 info@powrmatic.co.uk www.powrmatic.co.uk
Powermax	All enquiries to Baxi Potterton
Protherm	All enquiries to Hepworth Heating
Proscon	All enquiries to Heatrae-Sadia
Pullin Electronics	All enquiries to Heatrae-Sadia
Pyrocraft	All enquiries to Time and Temperature
Quick Spares Ltd	Storey Lane Industrial Estate, Rain Hill, Merseyside, L35 9LZ Tel: 0151 426 1393; Fax: 0151 426 1384
Radiant Boilers (Agent)	Unit 16, Spurlings Yard, Wallington, Gareham, Hampshire, PO17 6AB Tel: 01329 828555; Fax: 01329 823208
Radiation	All enquiries to Hepworth Heating Ltd
Randall Electronics	All enquiries to Danfoss Randall
Ravenheat	Chartists Way, Morley, Leeds, LS27 9ET Tel: 0113 252 7007; Fax: 0113 238 0229 enquiries@ravenheat.co.uk www.ravenheat.co.uk
Rayburn	All enquiries to Aga-Rayburn

Redring Electric Ltd	See Applied Energy
Riello Ltd	The Ermine Centre, Ermine Business Park, Huntingdon, Cambs, PE18 6XX Tel: 01480 432144; Fax: 01480 432191 info@rielloburners.co.uk www.rielloburners.co.uk
Ringdale UK Ltd	56 Victoria Road, Burgess Hill, West Sussex, RH15 9LR 01444 871349; Fax: 01444 870228 sales@ringdale.com www.ringdale.com
Saacke Ltd	Marshlands Spur, Farlington, Portsmouth, Hants, PO6 1RX Tel: 02392 383111; Fax: 02392 327120 admin@saacke.co.uk www.saacke.co.uk
Sangamo	Industrial Estate, Port Glasgow, Renfrewshire, PA14 5XG Tel: 01475 745131; Fax: 01475 744567 enquiries@sangamo.co.uk www.sangamo.co.uk
Satchwell Control Systems	All enquiries to Invensys
Satchwell Sunvic	All enquiries to Pegler or Sunvic
Saunier Duval	All enquiries to Hepworth Heating
Sauter Automation	Inova House, Hampshire International Business Park, Crockford Lane, Chineham, Basingstoke, RG24 8WH Tel: 01256 374400; Fax: 01256 374455 info@uk.sauter-bc.com www.sauter-controls.com
Servowarm	9 The Gateway Centre, Coronation Road, Cressex Business Park, High Wycombe, Bucks, HP12 3SU Tel: 0845 600 2266; Fax: 01494 492321 sales@servowarm.co.uk www.servowarm.co.uk
Siebe	All enquiries to Invensys
Siemens HVAC Products	Hawthorne Road, Staines, Middlesex, TW18 2AY Tel: 01527 406224; Fax: 01527 406207 www.landisstaefa.co.uk
Sime Ltd	Unit 2, Enterprise Way, Bradford Road, Idle, Bradford, BD10 8EW Tel: 0870 9911114; Fax: 0870 9911115 enquiries@sime.ltd.uk www.sime.ltd.uk
Sinclair	All enquiries to Halstead Heating
Smith Meters	All enquiries to Invensys
Smiths Industries	All enquiries to Timeguard
Sopac-Jaeger Controls	17 Invincible Road, Farnborough, Hants, GU14 7QN Tel: 01252 511981; Fax: 01252 524018 jaegercontrols@btconnect.com
Southern Digital	No longer trading
Stelrad-Ideal	All enquiries to Caradon-Ideal
Strebel Ltd	1F Albany Park Industrial Estate, Frimley Road, Camberley, Surrey, GU16 7PB Tel: 01276 685422; Fax: 01276 685405 info@strebel.co.uk www.strebel.co.uk
Sunvic Controls	Bellshill Road, Uddington, Glasgow, G71 6NP Tel: 01698 812944; Fax: 01698 813637 sales@sunvic.co.uk www.sunvic.co.uk
Superswitch Electric	Novar ED&S, The Arnold Centre, Paycocke Road, Basildon, Essex, SS14 3EA. Tel: 01268 563000 www.ade.co.uk
Switchmaster Controls	All enquiries to Invensys

Teddington Controls	Holmbush, St Austell, Cornwall, PL25 3HG Tel: 01726 222522; Fax: 01726 222504 gas@tedcon.com www.tedcon.com
Thermecon Boilers	All enquiries to GAH Heating Products
Thorn Heating	All enquiries to Baxi Potterton
TI Glow-Worm	All enquiries to Hepworth Heating
TI Radiation	All enquiries to Hepworth Heating
Time & Temperature	Unit 56, Plume Street Industrial Estate, Plume Street, Aston, Birmingham, B6 7RT Tel: 0121 327 2717; Fax: 0121 327 2395 www.plumbbase.com
Timeguard	Waterloo Road, London, NW2 7UR Tel: 020 8450 8944; Fax: 020 8452 5143 csc@timeguard.com www.timeguard.com
Tower	See Grasslin (UK) Ltd
Trac Time Controls	Thirsk Industrial Estate, Thirsk, North Yorks, YO7 3BX Tel: 01845 526006; Fax: 01845 526010 sales@trac.co.uk www.trac.co.uk
Trianco Redfyre Ltd	Thorncliffe, Chapeltown, Sheffield, S35 2PZ Tel: 0114 257 2300; Fax: 0114 245 3021 info@trianco.co.uk www.trianco.co.uk
Tricom Group Ltd	All enquiries to Hepworth Heating
Trisave	All enquiries to Crosslee PLC
Turkington Engineering	Unit 12, Whitegate Industrial Estate, Wrexham, LL13 8UG Tel: 01978 363048; Fax: 01978 363046 Eurocal@wrexham.plus.com www.eurocalboilers.com
Unidare	See Dimplex
Vaillant Ltd	Valliant House, Medway City Estate, Trident Close, Rochester, Kent, ME2 4EZ Tel: 01634 292300; Fax: 01634 290166 info@vaillant.co.uk www.vaillant.co.uk
Venner	All enquiries to Heating Controls Services
Viessmann Ltd	Hortonwood 32, Telford, Shropshire, TF1 4EU Tel: 01952 675000; Fax: 01952 675040 Info-uk@viessmann.com www.viessmann.com
Vokera Ltd	4th Floor, Catherine House, Boundary Way, Hemel Hempstead, Herts, HP2 7RP Tel: 0870 333 0220; Fax: 01442 281460 enquiries@vokera.co.uk www.vokera.co.uk
Warmflow Engineering	144 Bradford Road, Manchester, M40 7AS Tel: 0161 205 4202; Fax: 0161 205 4818 sales@warmflowgb.co.uk www.warmflowboilers.co.uk
WarmWorld UK Ltd	1 Hanham Business Park, Memorial Road, Hanham, Bristol, BS15 3JE Tel: 0117 949 8800; Fax: 0117 949 8888 dataterm@warmworld.co.uk www.warmworld.co.uk
Wickes (Boilers)	All enquiries to Halstead Heating
Wickes (Controls)	All enquiries to Siemens

Wilo Samson Pumps
Centrum 100, Second Avenue, Burton on Trent, DE14 2WJ
Tel: 01283 523000; Fax: 01283 523099
sales@wilo.co.uk www.wilo.co.uk

Worcester Heat Systems
Cotswold Way, Warndon, Worcester, WR4 9SW
Tel: 01905 754624; Fax: 01905 754619
www.worcester-bosch.co.uk

Yorkpark Ltd
All enquiries to Carfield, Unit B, Blenheim House, 1 Blenheim Road, Epsom, Surrey, KT19 9AP
Tel: 01372 722277

INDEX